Differential Equations

Differential Equations

Sofia Lynch

Larsen & Keller
www.larsen-keller.com

Differential Equations
Sofia Lynch
ISBN: 978-1-64172-680-1 (Hardback)

Larsen & Keller

Published by Larsen and Keller Education,
5 Penn Plaza,
19th Floor,
New York, NY 10001, USA

Cataloging-in-Publication Data

Differential equations / Sofia Lynch.
    p. cm.
Includes bibliographical references and index.
ISBN 978-1-64172-680-1
1. Differential equations. 2. Calculus. I. Lynch, Sofia.
QA371 .D54 2022
515.35--dc23

For more information regarding Larsen and Keller Education and its products, please visit the publisher's website www.larsen-keller.com

Table of Contents

Preface   **VII**

Chapter 1 **Differential Equations: An Introduction**   **1**
    a. Differentiation   1
    b. Partial Differentiation   4
    c. Integration   15
    d. Differential Equation   24
    e. Order and Degree of Differential Equation   30

Chapter 2 **Ordinary Differential Equations of First Order and First Degree**   **33**
    a. Ordinary Differential Equations   33
    b. Differential Equations of First Order and First Degree   38
    c. Separable Equations   39
    d. Homogeneous Equations   50
    e. Exact Differential Equations   55
    f. First Order Linear Differential Equations   72
    g. Bernoulli Differential Equation   83

Chapter 3 **Linear Differential Equations of Second Order and Higher Order**   **90**
    a. Second Order Linear Homogeneous Differential Equations with Variable Coefficients   90
    b. Second Order Linear Homogeneous Differential Equations with Constant Coefficients   97
    c. Second Order Linear Non-homogeneous Differential Equations   109
    d. Higher Order Linear Homogeneous Differential Equations with Constant Coefficients   132
    e. Higher Order Linear Homogeneous Differential Equations with Variable Coefficients   136
    f. Higher Order Linear Non-homogeneous Differential Equations with Constant Coefficients   141
    g. System of Simultaneous Linear Differential Equations with Constant Coefficients   151

Chapter 4 **Partial Differential Equations**   **157**
    a. First Order Partial Differential Equations   159
    b. First Order Linear Partial Differential Equations   170
    c. First Order Non-linear Partial Differential Equations   176
    d. Non-homogeneous Linear Partial Differential Equations   185
    e. Second Order Partial Differential Equations   188
    f. Higher-Order Partial Differential Equations   199
    g. The Heat Equation   201

Chapter 5   **Numerical Solution of Differential Equations**   **207**
     a. Picard's Method   207
     b. Euler's Method   212
     c. Runge-Kutta Methods   218

Permissions

Index

Preface

A mathematical equation which relates some function with its derivatives is known as a differential equation. While applying a differential equation, the physical quantities are represented by functions and the rates of change are represented by derivatives. The relationship between the two is defined by a differential equation. There are several types of differential equations such as ordinary differential equations, partial differential equations and non-linear differential equations. Ordinary differential equations contain an unknown function of one real or complex variable x, its derivatives, and some given functions of x. A differential equation which contains unknown multivariable functions and their partial derivatives is known as a partial differential equation. A differential equation which is not a linear equation in the unknown function and its derivatives is known as a non-linear differential equation. This book provides comprehensive insights into the field of differential equations. Some of the diverse topics covered herein address the varied branches that fall under this category. Those with an interest in this field would find this book helpful.

A detailed account of the significant topics covered in this book is provided below:

Chapter 1- Differential equation can be termed as a relationship between the derivatives of some functions. The degree of a differential equation is the highest power of the order derivative and the order of a differential equation refers to the highest number of derivatives in an equation. This chapter has been carefully written to provide an easy understanding of differential equations.

Chapter 2- Ordinary differential equations of first order include a(x) and f(x) as the continuous functions of x. The two methods of solving such differential equations are method of integrating factor and method of variation of a constant. This chapter discusses in detail about the ordinary differential equations of first order and first degree.

Chapter 3- Any differential equation which is defined by a linear polynomial in the unknown function is known as a linear differential equation. This chapter delves into the linear differential equations of the second order and higher order to provide a holistic understanding of differential equations. The topics elaborated in this chapter will help in gaining a better perspective about linear differential equations of second order and higher order.

Chapter 4- Partial differentiation equations comprise of an unknown variable and their partial derivatives. This kind of equations can be used in different phenomena like sound, heat, diffusion, fluid dynamics, gravitation, etc. All the aspects related to partial differential equations have been carefully analyzed in this chapter.

Chapter 5- Numerical solution of differential equations are used in finding numerical approximations to the solutions of ordinary differential equations. It uses Picard's method, Euler method and Runge-Kutta method to find these approximations. This chapter closely examines these methods of numerical solutions of differential equations to provide an extensive understanding of the subject.

It gives me an immense pleasure to thank our entire team for their efforts. Finally in the end, I would like to thank my family and colleagues who have been a great source of inspiration and support.

Sofia Lynch

Differential Equations: An Introduction

Differential equation can be termed as a relationship between the derivatives of some functions. The degree of a differential equation is the highest power of the order derivative and the order of a differential equation refers to the highest number of derivatives in an equation. This chapter has been carefully written to provide an easy understanding of differential equations.

Differentiation

In mathematics, differentiation is the process of finding the derivative, or rate of change, of a function. In contrast to the abstract nature of the theory behind it, the practical technique of differentiation can be carried out by purely algebraic manipulations, using three basic derivatives, four rules of operation, and a knowledge of how to manipulate functions.

The three basic derivatives (D) are: (1) for algebraic functions, $D(x^n) = nx^{n-1}$, in which n is any real number; (2) for trigonometric functions, $D(\sin x) = \cos x$; and (3) for exponential functions, $D(e^x) = e^x$.

For functions built up of combinations of these classes of functions, the theory provides the following basic rules for differentiating the sum, product, or quotient of any two functions $f(x)$ and $g(x)$ the derivatives of which are known (where a and b are constants): $D(af + bg) = aDf + bDg$ (sums); $D(fg) = fDg + gDf$ (products); and $D(f/g) = (gDf - fDg)/g^2$ (quotients).

The other basic rule, called the chain rule, provides a way to differentiate a composite function. If $f(x)$ and $g(x)$ are two functions, the composite function $f(g(x))$ is calculated for a value of x by first evaluating $g(x)$ and then evaluating the function f at this value of $g(x)$; for instance, if $f(x) = \sin x$ and $g(x) = x^2$, then $f(g(x)) = \sin x^2$, while $g\left(f\left(x\right)\right) = \left(\sin x\right)^2$. The chain rule states that the derivative of a composite function is given by a product, as $D(f(g(x))) = Df(g(x)) \cdot Dg(x)$. In words, the first factor on the right, $Df(g(x))$, indicates that the derivative of $Df(x)$ is first found as usual, and then x, wherever it occurs, is replaced by the function $g(x)$. In the example of $\sin x^2$, the rule gives the result $D(\sin x^2) = D\sin(x^2) \cdot D(x^2) = (\cos x^2) \cdot 2x$.

In the German mathematician Gottfried Wilhelm Leibniz's notation, which uses d/dx in place of D and thus allows differentiation with respect to different variables to be made explicit, the chain rule takes the more memorable "symbolic cancellation" form:

$$d\left(f\left(g\left(x\right)\right)\right)/dx = df/dg \cdot dg/dx.$$

The Derivative

Consider a function y = f(x). For some point x, we can find,

- The slope of the tangent to the curve described by f(x).

- The instantaneous rate at which y is changing.

By the following method: Find the slope of the line segment joining $(x, f(x))$ and a nearby point $(x + h, f(x + h))$ as shown below:

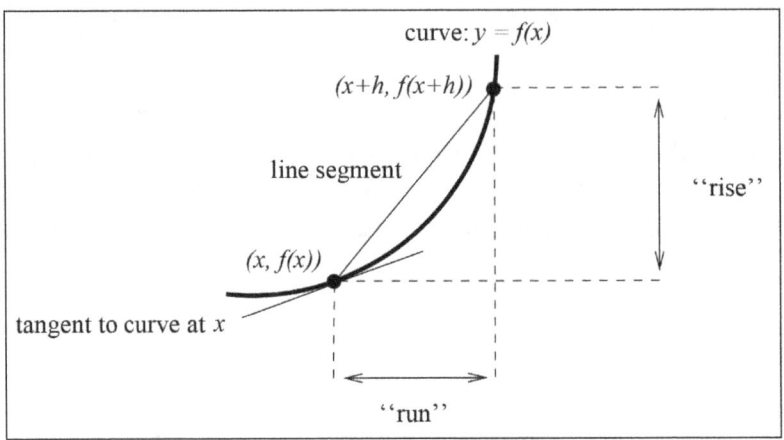

So,

$$\frac{\text{"rise"}}{\text{"run"}} = \frac{\Delta y}{\Delta x} = \frac{f(x+h) - f(x)}{x+h-x}$$

and, in the limit as $h \to 0$,

$$\text{slop of tangent to } f(x) \text{ at } x = \lim_{h \to 0} \frac{\Delta y}{\Delta x} = \lim_{h \to 0} \frac{f(x+h) - f(x)}{h}.$$

Finding derivatives this way is tedious but a number of shortcut rules are available. In fact we can use these rules to find the function that gives the slope of the tangent to f(x) at any point x. This derivative function is given the name $\frac{dy}{dx}$ or $f'(x)$.

$\frac{dy}{dx}$ or $f'(x)$.

$f(x)$	$f'(x)$
k (a constant)	0
x^n	nx^{n-1}, for all n
e^x	e^x
In x	$\frac{i}{x}$
$kf(x)$	$kf'(x)$

$f(x)+g(x)$	$f'(x)+g'(x)$
Product Rule $f(x)g(x)$	$f'(x)g(x)+f(x)g'(x)$
Quotient Rule $\dfrac{f(x)}{g(x)}$	$\dfrac{f'(x)g(x)-f(x)g'(x)}{[g(x)]^2}$

A constant doesn't change.

Eg. $f(x)=x^{-3}$ so $f'(x)=-3x^{-4}$.

Eg. $y=2x^2$ so $\dfrac{dy}{dx}=2\times 2x^1=4x$.

Eg. $f(x)=x^3+\ln x$ so $f'(x)=3x^2+\dfrac{1}{x}$.

Example: Differentiate $f(x)=2x+3$.

Solution: $f'(x)=2\times 1x^{1-1}+0=2x^0=2$. (This should not be a surprise since $f(x)$ is clearly a straight line with slope 2. The solution $f'(x)=2$ indicates that the tangent has slope 2 for any value of x, as required).

Example: Differentiate $f(x)=4x^3\ln x$.

Solution: This is a product of x³ and ln x with a constant multiple of 4. So,

$$f'(x)=4\left(3x^2\times \ln x+x^3\times \dfrac{1}{x}\right)$$
$$=12x^3\ln x+4x^2.$$

Example: Differentiate $y\,\dfrac{x+4}{2x+5}$.

Solution: This is a quotient of x + 4 and 2x + 5 so,

$$\dfrac{dy}{dx}=\dfrac{1\times(2x+5)-(x+4)\times 2}{(2x+5)^2}$$
$$=-\dfrac{3}{(2x+5)^2}.$$

The Chain Rule

This is the most useful rule of the lot and is based on the following idea:

$$\dfrac{dy}{dx}=\dfrac{dy}{du}\times \dfrac{du}{dx},$$

where, u is a function of x that suits you.

Example: $y = e^{3x}$ can't be differentiated by the current rules but it could be done if $u(x) = 3x$ and we apply the chain rule.

$$y = e^u \quad \text{so} \quad \frac{dy}{du} = e^u \quad \text{and} \quad \frac{dy}{dx} = 3.$$

Hence,

$$\frac{dy}{dx} = e^u \times 3 = 3e^{3x}.$$

Example: Consider $y = (2x+1)^3$. If we expand the brackets we get,

$$y = (4x^2 + 4x + 1)(2x + 1) = 8x^3 + 12x^2 + 6x + 1.$$

and hence,

$$\frac{dy}{dx} = 24x^2 + 24x + 6 = 6(4x^2 + 4x + 1) = 6(2x+1)^2.$$

The chain rule is more efficient (especially in cases where the power is higher than) if we let,

$$u(x) = 2x + 1.$$

$$y = u^3 \quad \text{so} \quad \frac{dy}{dx} = 3u^2 \quad \text{and} \quad \frac{dy}{dx} = 2.$$

Hence,

$$\frac{dy}{dx} = 3u^2 \times 2 = 6(2x+1)^2 \text{ as before}.$$

Partial Differentiation

Partial differentiation is used to differentiate mathematical functions having more than one variable in them. In ordinary differentiation, we find derivative with respect to one variable only, as function contains only one variable. So partial differentiation is more general than ordinary differentiation.

Partial Derivatives

For a function f(x) of a single variable the derivative of f at x = a,

$$f'(a) = \lim_{h \to 0} \frac{f(a+h) - f(a)}{h}.$$

is the instantaneous rate of change of f at a, and is equal to the slope of the tangent line to the graph of f(x) at (a, f(a)).

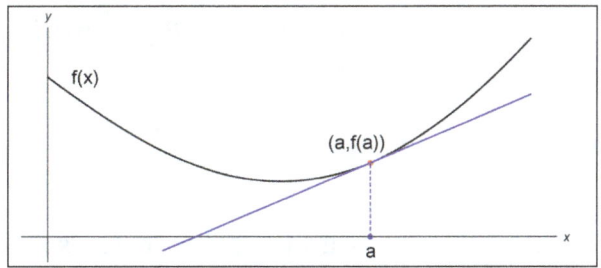

Equation of the tangent line: $y = f(a) + f'(a)(x-a)$.

Consider $f(x,y)$. If we fix $y = b$ where b is a number from the domain of f then $f(x,b)$ is a function of a single variable x and we can calculate its derivative at some $x = a$. This derivative is called the partial derivative of $f(x,y)$ with respect to x at (a, b) and is denoted by,

$$f_x(a,b) \text{ or by } \frac{\partial f(a,b)}{\partial x}$$

$$f_x(a,b) = \frac{\partial f(a,b)}{\partial x} = \frac{d}{dx}[f(x,b)]\Big|_{x=a} = \lim_{h \to 0} \frac{f(a+h,b) - f(a,b)}{h}.$$

If $f(x,y) = x$ then $\dfrac{\partial x}{\partial x} = 1$, and If $f(x,y) = y$ then $\dfrac{\partial y}{\partial x} = 0$.

Geometrically, given the surface $z = f(x,y)$, we consider its intersection with the plane $y = b$ which is a curve. This curve is the graph of the function $f(x,b)$ and therefore the partial derivative $f_x(a,b)$ is the slope of the tangent line to the curve at $(a,b,f(a,b))$.

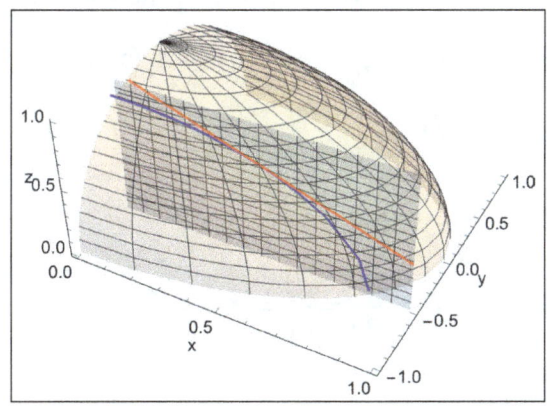

Equation of the tangent line: $x = t,\ y = b,\ z = f(a,b) + f_x(a,b)(t-a)$.

We call $f_x(a,b)$ the slope of the surface in the x-direction at (a, b).

Similarly, if we fix $x = a$ where a is a number from the domain of f then $f(a,y)$ is a function of a single variable y and we can calculate its derivative at some $y = b$. This derivative is called the partial derivative of $f(x,y)$ with respect to y at (a,b) and is denoted by,

$$f_y(a,b) \text{ or by } \frac{\partial f(a,b)}{\partial y},$$

$$f_y(a,b) = \frac{\partial f(a,b)}{\partial y} = \frac{d}{dy}[f(a,y)]\Big|_{y=b} = \lim_{h\to 0}\frac{f(a,b+h)-f(a,b)}{h},$$

if $f(x,y) = x$ then $\dfrac{\partial x}{\partial y} = 0$, and if $f(x,y) = y$ then $\dfrac{\partial y}{\partial y} = 1$.

The intersection of the surface $z = f(x,y)$, with the plane $x = a$ is a curve which is the graph of the function $f(a,y)$ and therefore the partial derivative $f_y(a,b)$ is the slope of the tangent line to the curve at $(a,b,f(a,b))$.

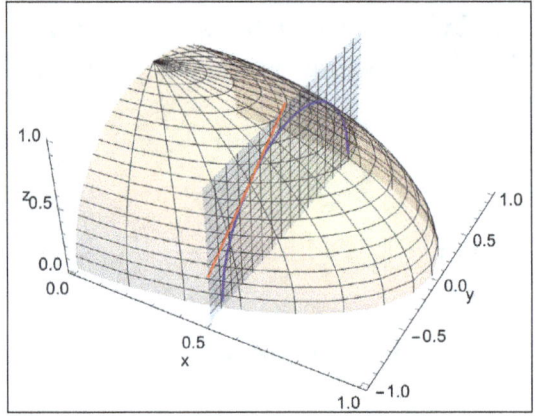

Equation of the tangent line: $x = a$, $y = t$, $z = f(a,b) + f_y(a,b)(t-a)$. We call $f_y(a,b)$ the slope of the surface in the y-direction at (a, b).

If we allow (a, b) to vary, the partial derivatives become functions of two variables:

$$a \to x, b \to y \text{ and } f_x(a,b) \to f_x(x,y), f_y(a,b) \to f_y(x,y)$$

$$f_x(x,y) = \lim_{h\to 0}\frac{f(x+h,y)-f(x,y)}{h}, \quad f_y(x,y) = \lim_{h\to 0}\frac{f(x,y+h)-f(x,y)}{h}.$$

Partial derivative notation: if z = f(x, y) then,

$$f_x = \frac{\partial f}{\partial x} = \frac{\partial z}{\partial x} = \partial_x f = \partial_x z, \quad f_y = \frac{\partial f}{\partial y} = \frac{\partial z}{\partial y} = \partial_y f = \partial_y z.$$

Second-order derivatives: $f_{xx}, f_{xy}, f_{yx}, f_{yy}$.

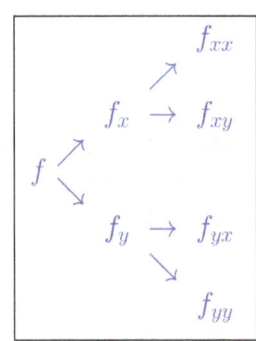

Notation

$$f_{xx} = \frac{\partial^2 f}{\partial x^2} = \frac{\partial}{\partial x}\left(\frac{\partial f}{\partial x}\right), \quad f_{xy} = \frac{\partial^2 f}{\partial y \partial x} = \frac{\partial}{\partial y}\left(\frac{\partial f}{\partial x}\right),$$

$$f_{yx} = \frac{\partial^2 f}{\partial x \partial y} = \frac{\partial}{\partial x}\left(\frac{\partial f}{\partial y}\right), \quad f_{yy} = \frac{\partial^2 f}{\partial y^2} = \frac{\partial}{\partial y}\left(\frac{\partial f}{\partial y}\right).$$

f_{xy} and f_{yx} are called the mixed second-order partial derivatives. f_x and f_y can be called first-order partial derivative.

Equality of Mixed Partial Derivatives

Theorem 1: Let f be a function of two variables. If f_{xy} and f_{yx} are continuous on some open disc, then $f_{xy} = f_{yx}$ on that disc.

Higher-order Derivatives

Third-order, fourth-order, and higher-order derivatives are obtained by successive differentiation.

$$f_{xxx} = \frac{\partial^3 f}{\partial x^3} = \frac{\partial}{\partial x}\left(\frac{\partial^2 f}{\partial x^2}\right), \quad f_{xyy} = \frac{\partial^3 f}{\partial y^2 \partial x} = \frac{\partial}{\partial y}\left(\frac{\partial^2 f}{\partial y \partial x}\right),$$

$$f_{xyxz} = \frac{\partial^4 f}{\partial z \partial x \partial y \partial x} = \frac{\partial}{\partial z}\left(\frac{\partial^3 f}{\partial x \partial y \partial x}\right).$$

For higher-order derivatives the equality of mixed partial derivatives also holds if the derivatives are continuous.

In what follows we always assume that the order of partial derivatives is irrelevant for functions of any number of independent variables.

Differentiability, Differentials and Local Linearity

For $f(x, y)$, the symbol Δf, called the increment of f, denotes the change,

$$\Delta f = f(a + \Delta x, b + \Delta y) - f(a, b).$$

For small $\Delta x, \Delta y$,

$$\Delta f \approx f_x(a,b)\Delta x + f_y(a,b)\Delta y.$$

A function $f(x, y)$ is said to be differentiable at (a, b) provided $f_x(a,b)$ and $f_y(a,b)$ both exist and,

$$\lim_{(\Delta x, \Delta y) \to (0,0)} \frac{\Delta f - f_x(a,b)\Delta x - f_y(a,b)\Delta y}{\sqrt{(\Delta x)^2 + (\Delta y)^2}} = 0$$

For $f(x, y, z)$,

$$\Delta f = f(a + \Delta x, b + \Delta y, c + \Delta z) - f(a, b, c).$$

For small $\Delta x, \Delta y, \Delta z$,

$$\Delta f \approx f_x(a,b,c)\Delta x + f_y(a,b,c)\Delta y + f_z(a,b,c)\Delta z.$$

and $f(x, y, z)$ z) is differentiable at (a, b, c) if,

$$\lim_{(\Delta x, \Delta y, \Delta z) \to (0,0,0)} \frac{\Delta f - f_x(a, b, c)\Delta x - f_y(a,b,c)\Delta y - f_z(a,b,c)\Delta z}{\sqrt{(\Delta x)^2 + (\Delta y)^2 + +(\Delta z)^2}} = 0.$$

Theorem 2: If a function is differentiable at a point, then it is continuous at that point.

Theorem 3: If all first-order derivatives of f exist and are continuous at a point, then f is differentiable at a point.

Differentials

If $z = f(x, y)$ is differentiable at (a, b) we let,

$$dz = f_x(a,b)dx + f_y(a,b)dy.$$

Denote a new function with dependent variable dz and independent variables dx, dy. It is called the total differential of z (or f) at (a, b). It is a linear function of dx and dy.

Note that $\Delta z \approx dz$ if $\Delta x = dx$ and $\Delta y = dy$.

If we allow (a, b) to vary, the differential becomes a function of four variables, dx, dy, x, y:

$$a \to x, b \to y \implies dz = f_x(x, y)dx + f_y(x, y)dy.$$

If f(x, y) is differentiable at (a, b) then,

$$L(x, y) = f(a,b) + f_x(a,b)(x-a) + f_y(a,b)(y-b).$$

It is called the local linear approximation of f at (a, b).

Its graph is the tangent plane to the surface $z = f(x, y)$ at $(a, b, f(a,b))$.

The Chain Rule

Recall,

$$y = f(x(t)) \implies \frac{dy}{dt} = \frac{dy}{dx}\frac{dx}{dt}.$$

Because,

$$\Delta y \approx \frac{dy}{dx}\Delta x, \quad \Delta x \approx \frac{dx}{dt}\Delta t.$$

Let $z = f(x, y)$ and $x = x(t)$, $y = y(t)$. Then $z = f(x(t), y(t))$ is a function of the single variable t.

$$\Delta z \approx \frac{\partial z}{\partial x}\Delta x + \frac{\partial z}{\partial y}\Delta y, \quad \Delta x \approx \frac{dx}{dt}\Delta t, \quad \Delta y \approx \frac{dy}{dt}\Delta t,$$

and therefore,

$$\frac{dz}{dt} = \frac{\partial z}{\partial x}\frac{dx}{dt} + \frac{\partial z}{\partial y}\frac{dy}{dt}.$$

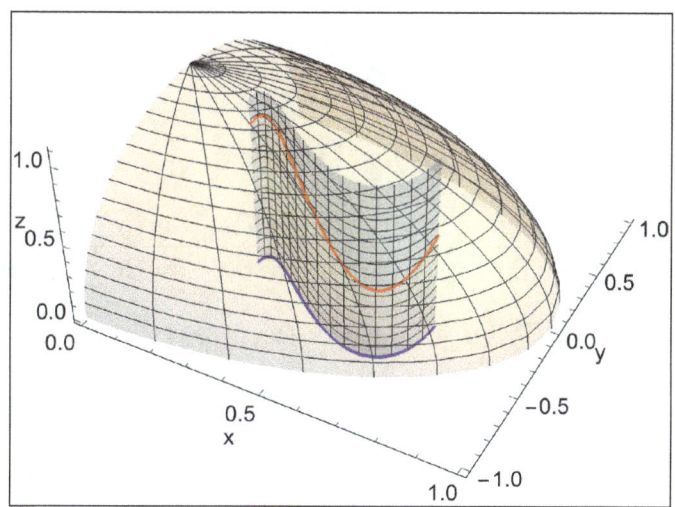

Implicit Differentiation

Let $z = f(x, y)$ and $y = y(x)$. Then,

$$\frac{dz}{dx} = \frac{\partial f}{\partial x}\frac{dx}{dx} + \frac{\partial f}{\partial y}\frac{dy}{dx} = \frac{\partial f}{\partial x} + \frac{\partial f}{\partial y}\frac{dy}{dx}.$$

Suppose $y(x)$ is such that $f(x, y(x)) = \text{const}$. Then $\dfrac{dz}{dx} = 0$ and,

$$\frac{\partial f}{dx} + \frac{\partial f}{\partial y}\frac{dy}{dx} = 0 \Rightarrow \frac{dy}{dx} = -\frac{f_x}{f_y} \quad \text{if} \quad f_y \neq 0.$$

The Chain Rule for Partial Derivatives

Let $y = f(x)$ and $x = x(u, v)$.

Then $y = f(x(u, v))$ is a function of u and v, and,

$$\Delta y \approx \frac{dy}{dx}\Delta x, \quad \Delta x \approx \frac{\partial x}{\partial u}\Delta u + \frac{\partial x}{\partial v}\Delta v.$$

Thus,

$$\frac{\partial y}{\partial u} = \frac{dy}{dx}\frac{\partial x}{\partial u}, \quad \frac{\partial y}{\partial v} = \frac{dy}{dx}\frac{\partial x}{\partial v}.$$

Let $z = f(x, y)$ and $x = x(u, v)$, $y = y(u, v)$.

Then $x = f(x(u, v), y(u, v)$ is a function of u and v, and,

$$\Delta z \approx \frac{\partial z}{\partial x}\Delta x + \frac{\partial z}{\partial y}\Delta y, \quad \Delta x \approx \frac{\partial x}{\partial u}\Delta u + \frac{\partial x}{\partial v}\Delta v, \quad \Delta y \approx \frac{\partial y}{\partial u}\Delta u + \frac{\partial y}{\partial v}\Delta v.$$

Thus,

$$\frac{\partial z}{\partial u} = \frac{\partial z}{\partial x}\frac{\partial x}{\partial u} + \frac{\partial z}{\partial y}\frac{\partial y}{\partial u}, \qquad \frac{\partial z}{\partial v} = \frac{\partial z}{\partial x}\frac{\partial x}{\partial v} + \frac{\partial z}{\partial y}\frac{\partial y}{\partial v}.$$

Let, $w = f(x, y, z)$ and $x = x(u, v)$, $y = y(u, v)$, $z = z(u, v)$.

$$\frac{\partial w}{\partial u} = \frac{\partial w}{\partial x}\frac{\partial x}{\partial u} + \frac{\partial w}{\partial y}\frac{\partial y}{\partial u} + \frac{\partial w}{\partial z}\frac{\partial z}{\partial u}, \qquad \frac{\partial w}{\partial v} = \frac{\partial w}{\partial x}\frac{\partial x}{\partial v} + \frac{\partial w}{\partial y}\frac{\partial y}{\partial v} + \frac{\partial w}{\partial z}\frac{\partial z}{\partial v}$$

Let, $w = f(x_1, ..., x_n)$ and $x_1 = x_1(u_1, ..., u_m), ..., x_n = x_n(u_1, ..., u_m)$.

$$\frac{\partial w}{\partial u_\alpha} = \sum_{i=1}^{n} \frac{\partial w}{\partial x_i}\frac{\partial x_i}{\partial u_\alpha}, \qquad \alpha = 1, ..., m.$$

Directional Derivatives and Gradients

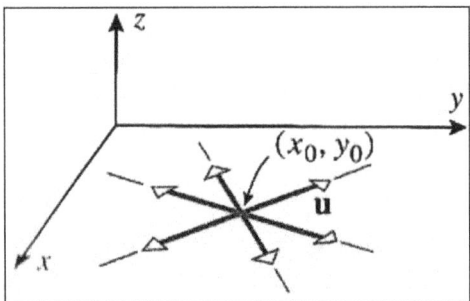

Suppose we need to compute the rate of change of $f(x, y)$ with respect to the distance from a point (a, b) in some direction. Let $\vec{u} = u_1\vec{i} + u_2\vec{j}$ be the unit vector that has its initial point at (a, b) and points in the desired direction. It determines a line in the xy-plane:

$$x = a + s u_1, \quad y = b + s u_2$$

where s is the arc length parameter that has its reference point at (a, b) and has positive values in the direction of $\vec{u}$.

The directional derivative of $f(x, y)$ in the direction of $\vec{u}$ at (a, b) is denoted by $D_{\vec{u}}f(a, b)$ and is defined by,

$$D_{\vec{u}}f(a, b) = \frac{d}{ds}[f(a + s u_1, b + s u_2)]\Big|_{s=0} = f_x(a, b)u_1 + f_y(a, b)u_2$$

provided this derivative exists.

Analytically, $D_{\vec{u}}f(a,b)$ is the instantaneous rate of change of $f(x,y)$ with respect to the distance in the direction of $\vec{u}$ at the point (a, b).

Geometrically, $D_{\vec{u}}f(a,b)$ is the slope of the surface $z = f(x,y)$ in the direction of $\vec{u}$ at the point $\left(a,\,b,\,f\left(a,\,b\right)\right)$.

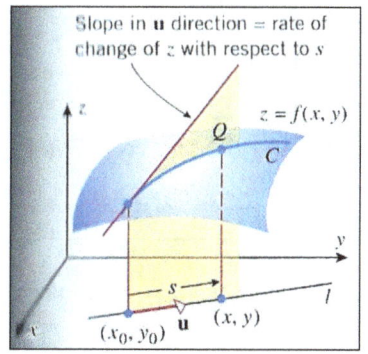
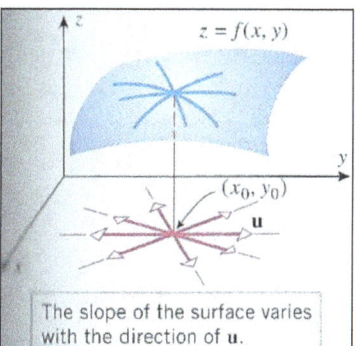

Generalisation to $f(x,y,z)$ (and $f(x_1,....,x_n)$) is straightforward.

Let $\vec{u} = u_1\vec{i} + u_2\vec{j} + u_3\vec{k}$ be a unit vector.

The directional derivative of $f(x,y,z)$ in the direction of $\vec{u}$ at (a, b, c) is denoted by $D_{\vec{u}}f(a,b,c)$ and is defined by,

$$D_{\vec{u}}f(a,b,c) = \frac{d}{ds}[f(a+su_1,\ b+su_2,\ c+su_3)]\big|_{s=0}$$
$$= f_x(a,b,c)u_1 + f_y(a,b,c)u_2 + f_z(a,b,c)u_3.$$

The Gradient

Note that,

$$D_{\vec{u}}f = f_x u_1 + f_y u_2 + f_z u_3 = (f_x\vec{i} + f_y\vec{j} + f_z\vec{k})\cdot(u_1\vec{i} + u_2\vec{j} + u_3\vec{k}).$$

Let, $\vec{e}_i$ be the standard orthonormal coordinate basis of $\mathbb{R}^n$, so that $\vec{r} = \sum_{i=1}^{n} x_i\vec{e}_i$.

The gradient of $f(x,\ ,x)$ is defined by,

$$\vec{\nabla}f(x_1,\cdots,x_n) = \sum_{i=1}^{n} \frac{\partial f(x_1,\cdots,x_n)}{\partial x_i}\vec{e}_i.$$

In particular,

$$\vec{\nabla}f(x,y) = f_x(x,y)\vec{i} + f_y(x,y)\vec{j},$$

$$\vec{\nabla}f(x,y,z) = f_x(x,y,z)\vec{i} + f_y(x,y,z)\vec{j} + f_z(x,y,z)\vec{k}.$$

The symbol $\vec{\nabla}$ is read as either "nabla" (from ancient Hebrew) or "del" (it is inverted Δ).

$$D_{\vec{u}}f(a,b) = \vec{\nabla}f(a,b)\cdot\vec{u},\ \ D_{\vec{u}}f(a,b,c) = \vec{\nabla}f(a,b,c)\cdot\vec{u},\ \ D_{\vec{u}}f = \vec{\nabla}f\cdot\vec{u}.$$

Properties of the Gradient

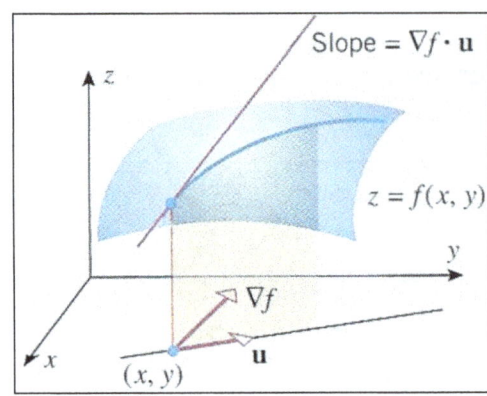 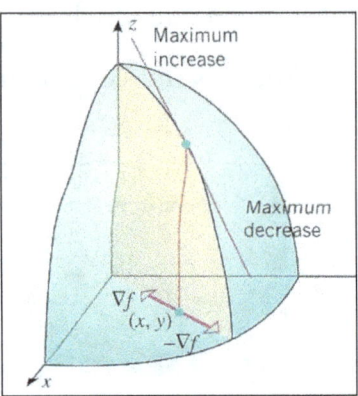

$$D_{\vec{u}}f(a,b) = \vec{\nabla}f(a,b)\cdot\vec{u} = \left|\vec{\nabla}f(a,b)\right|\left|\vec{u}\right|\cos\theta = \left|\vec{\nabla}f(a,b)\right|\cos\theta.$$

Since $-1 \le \cos\theta \le 1$, if $\left|\vec{\nabla}f(a,b)\right| \ne 0$ then the maximum value of $D_{\vec{u}}f(a,b)$ is $\left|\vec{\nabla}f(a,b)\right|$ and it occurs when $\theta = 0$ that is, when $\vec{u}$ is in the direction of $\vec{\nabla}f(a,b)$.

Geometrically, the maximum slope of the surface $z = f(x,y)$ at (a, b) is in the direction of the gradient and is equal to $\left|\vec{\nabla}f(a,b)\right|$.

If $\left|\vec{\nabla}f(a,b)\right| = 0$ then $D_{\vec{u}}f(a,b) = 0$ in all directions at (a, b). It occurs where the surface $z = f(x,y)$ has a relative maximum or minimum or a saddle point.

Since $D_{\vec{u}}f(x_1,\dots,x_n) = \left|\vec{\nabla}f(x_1,\dots,x_n)\right|\cos\theta,$ these properties hold for functions of any number of variables.

Theorem 4: Let f be a function differentiable at a point P.

1. If $\vec{\nabla}f = \vec{0}$ at P then all directional derivatives of f at P are 0.

2. If $\vec{\nabla}f \ne \vec{0}$ at p then the derivative in the direction of $\vec{\nabla}f$ at P has the largest value equal to $\left|\vec{\nabla}f\right|$ at P.

3. If $\vec{\nabla}f \ne \vec{0}$ at P then the derivative in the direction opposite to that of $\vec{\nabla}f$ at P has the smallest value equal to $-\left|\vec{\nabla}f\right|$ at P.

Gradients are Normal to Level Curves and Level Surfaces

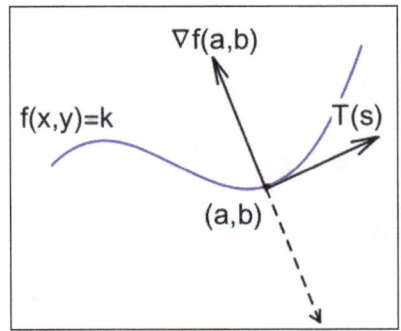

Level curve C: $f(x, y) = k$.

Let, C be smoothly parametrised as $x = x(s)$, $y = y(s)$ where s is an arc length parameter. The unit tangent vector to C is,

$$\vec{T}(s)\frac{dx}{ds}\vec{i} + \frac{dy}{ds}\vec{j}.$$

Since $f(x, y)$ is constant on C we expect $D_{\vec{T}}f(x, y) = 0$ indeed,

$$D_{\vec{T}}f(x, y) = \vec{\nabla}f \cdot \vec{T} = (f_x\vec{i} + f_y\vec{j}) \cdot (\frac{dx}{ds}\vec{i} + \frac{dy}{ds}\vec{j})$$

$$= f_x\frac{dx}{ds} + f_y\frac{dy}{ds} = \frac{d}{ds}f(x(s), y(s)) = 0 \Rightarrow \vec{\nabla}f \perp \vec{T}.$$

Thus if (a, b) belongs to the level curve, and $\vec{\nabla}f(a,b) \neq \vec{0}$ then $\vec{\nabla}f(a,b)$ is normal to $\vec{T}$ at (a, b) and therefore to the level curve.

A vector is called normal to a surface at (a, b, c) if it is normal to a tangent vector to any curve on the surface through (a, b, c).

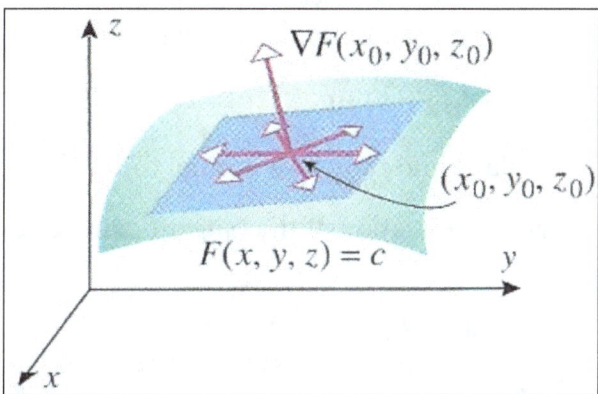

Level surface σ: F(x, y, z) = k.

Let C, smoothly parametrised as x = x(s), y = y(s), z = z(s) be any curve on σ through (a, b, c). The unit tangent vector to C is,

$$\vec{T}(s)\frac{dx}{ds}\vec{i} + \frac{dy}{ds}\vec{j} + \frac{dz}{ds}\vec{k}$$

and $D_{\vec{T}}F(x, y, z)$ is,

$$D_{\vec{T}}F(x, y, z) = \vec{\nabla}F \cdot \vec{T} = (F_x\vec{i} + F_y\vec{j} + F_z\vec{k}) \cdot (\frac{dx}{ds}\vec{i} + \frac{dy}{ds}\vec{j} + \frac{dz}{ds}\vec{k})$$

$$= F_x\frac{dx}{ds} + F_y\frac{dy}{ds} + F_z\frac{dz}{ds} = \frac{d}{ds}F(x(s), y(s), z(s)) = 0 \Rightarrow \vec{\nabla}F \perp \vec{T}.$$

Thus, $\vec{\nabla}F(a,b,c)$ is normal to $\vec{T}$ at (a, b, c) and therefore to σ.

Tangent Planes

Consider a level surface σ: F(x, y, z) = k, and let P = (a, b, c) belong to σ.

Since $\vec{\nabla}F(a,b,c)$ is normal to tangent vectors to curves on σ through P, all these tangent vectors belong to one and the same plane. This plane is called the tangent plane to the surface σ at P.

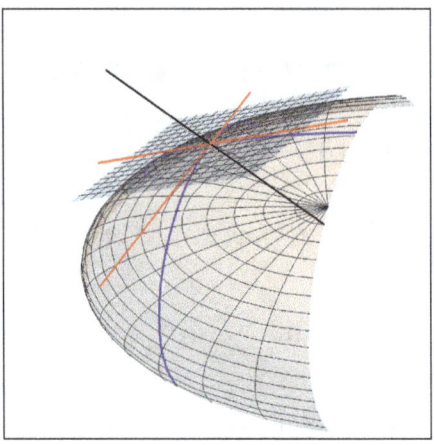

To find an equation of the tangent plane we use that if we know a vector ~n normal to a plane through a point $\vec{r}_0 = a\vec{i} + b\vec{j} + c\vec{k}$ then an equation of the plane is,

$$\vec{n}\cdot(\vec{r}-\vec{r}_0)=0 \Leftrightarrow n_1(x-a)+n_2(y-b)+n_3(z-c)=0.$$

Because $\vec{r}-\vec{r}_0$ is parallel to the plane and therefore normal to $\vec{n}$.

Choosing $\vec{n}=\vec{\nabla}F(a,b,c)$, we get the equation of the tangent plane to the level surface σ at P = (a, b, c):

$$F_x(a,b,c)(x-a)+F_y(a,b,c)(y-b)+F_z(a,b,c)(z-c)=0.$$

The line through P parallel to $\vec{\nabla}F(a,b,c)$, is perpendicular to the tangent plane, and is called the normal line to the surface σ at P. Its parametric equations are,

$$x=a+F_x(a,b,c)t, \quad y=b+F_y(a,b,c)t, \quad z=c+F_z(a,b,c)t.$$

Tangent Planes to z = f(x, y)

The graph of a function z = f(x, y) can be thought of as the level surface of the function F(x, y, z) = f(x, y) − z with constant 0.

We find,

1. The gradient,

    $$\vec{\nabla}F(a,b,c)=f_x(a,b)\vec{i}+f_y(a,b)\vec{j}-\vec{k}, \quad c=f(a,b).$$

2. The equation of the tangent plane to the surface z = f(x, y) at (a, b, f(a, b)),

$$f_x(a,b,c)(x-a) + f_y(a,b)(y-b) - (z-c) = 0 \Rightarrow$$
$$z = f(a,b) + f_x(a,b)(x-a) + f_y(a,b)(y-b).$$

that is the local linear approximation of f at (a, b).

3. The parametric equations of the normal line to the surface,

$$z = f(x,y) \text{ at } (a,b,f(a,b))$$
$$x = a + f_x(a,b)t, \quad y = b + f_y(a,b)t, \quad z = f(a,b) - t.$$

Integration

In mathematics, integration is a technique of finding a function $g(x)$ the derivative of which, $Dg(x)$, is equal to a given function $f(x)$. This is indicated by the integral sign "$\int$," as in $\int f(x)$, usually called the indefinite integral of the function. The symbol dx represents an infinitesimal displacement along x; thus $\int f(x)dx$ is the summation of the product of $f(x)$ and dx. The definite integral, written:

$$\int_a^b f(x)dx$$

with a and b called the limits of integration, is equal to $g(b) - g(a)$, where $Dg(x) = f(x)$.

Some antiderivatives can be calculated by merely recalling which function has a given derivative, but the techniques of integration mostly involve classifying the functions according to which types of manipulations will change the function into a form the antiderivative of which can be more easily recognized. For example, if one is familiar with derivatives, the function $1/(x+1)$ can be easily recognized as the derivative of $\log_e(x+1)$. The antiderivative of $(x^2 + x + 1)/(x+1)$ cannot be so easily recognized, but if written as $x(x+1)/(x+1) + 1/(x+1) = x + 1/(x+1)$, it then can be recognized as the derivative of $x^2/2 + \log_e(x+1)$. One useful aid for integration is the theorem known as integration by parts. In symbols, the rule is $\int f\, Dg = fg - \int gD f$. That is, if a function is the product of two other functions, f and one that can be recognized as the derivative of some function g, then the original problem can be solved if one can integrate the product gDf. For example, if $f = x$, and $Dg = \cos x$, then $\int x \cdot \cos x = x \cdot \sin x - \int \sin x = x \cdot \sin x - \cos x + C$. Integrals are used to evaluate such quantities as area, volume, work, and, in general, any quantity that can be interpreted as the area under a curve.

Integral

Definition of the Integral as an Anti-Derivative

If $\dfrac{d}{dx}(F(x)) = f(x)$ then $f(x)dx = F(x)$.

In words,

If the derivative of F(x) is f(x), then we say that an indefinite integral of f(x) with respect to x is F(x).

For example, since the derivative (with respect to x) of x^2 is 2x, we can say that an indefinite integral of 2x is x^2.

In symbols:

$$\frac{d}{dx}(x^2) = 2x, \quad \text{so} \quad \int 2x\,dx = x^2.$$

We say an indefinite integral, not the indefinite integral. This is because the indefinite integral is not unique. In our example, notice that the derivative of $x^2 + 3$ is also 2x, so $x^2 + 3$ is another indefinite integral of 2x. In fact, if c is any constant, the derivative of $x^2 + c$ is 2x and so $x^2 + c$ is an indefinite integral of 2x.

We express this in symbols by writing,

$$\int 2x\,dx = x^2 + c$$

where c is what we call an "arbitrary constant". This means that c has no specified value, but can be given any value we like in a particular problem. In this way we encapsulate all possible solutions to the problem of finding an indefinite integral of 2x in a single expression.

In most cases, if you are asked to find an indefinite integral of a function, it is not necessary to add the +c. However, there are cases in which it is essential. For example, if additional information is given and a specific function has to be found, or if the general solution of a differential equation is sought. So it is a good idea to get into the habit of adding the arbitrary constant every time, so that when it is really needed you don't forget it.

The inverse relationship between differentiation and integration means that, for every statement about differentiation, we can write down a corresponding statement about integration.

For example,

$$\frac{d}{dx}(x^4) = 4x^3, \quad \text{so} \quad \int 4x^3\,dx = x^4 + c.$$

Some Rules for Calculating Integrals

Rules for operating with integrals are derived from the rules for operating with derivatives. So, because,

$$\frac{d}{dx}(cf(x)) = c\frac{d}{dx}(f(x)), \text{ for any constant } c,$$

you have,

Rule 1:

$$\int (cf(x))\,dx = c\int (x)\,dx, \text{ for any constant } c.$$

For example $\int 10\cos x\,dx = 10\int \cos x\,dx = 10\sin x + c.$

It sometimes helps people to understand and remember rules like this if they say them in words. The rule given above says: The integral of a constant multiple of a function is a constant multiple of the integral of the function. Another way of putting it is You can move a constant past the integral sign without changing the value of the expression.

Similarly, from

$$\frac{d}{dx}(f(x)+g(x)) = \frac{d}{dx}(f(x)) + \frac{d}{dx}(g(x)),$$

you can derive the rule.

Rule 2:

$$\int (f(x)+g(x))dx = \int f(x)dx + \int g(x)dx.$$

For example,

$$\int \left(e^x + 2x\right)dx = \int e^x dx + \int 2x dx$$
$$= e^z + x^2 + c.$$

In words, the integral of the sum of two functions is the sum of their integrals. You can easily extend this rule to include differences as well as sums and to the case where there are more than two terms in the sum.

Examples: Find the following indefinite integrals:

$$\int (1+2x-3x^2+\sin x)dx$$

$$\int (3\cos x - \frac{1}{2}e^x)dx$$

Solutions:

$$\int (1+2x-3x^2+\sin x)dx = \int 1dx + \int 2x dx - \int 3x^2 dx - \int (-\sin x)dx$$
$$= x + x^2 - x^3 - \cos x + c.$$

You have written $+\int \sin x dx$ as $-\int(-\sin x)dx$ because $(-\sin x)$ is the derivative of cos x.

$$\int (3\cos x - \frac{1}{2}e^x)dx = 3\int \cos x dx - \frac{1}{2}\int e^x dx$$
$$= 3\sin x \frac{1}{2}e^x + c.$$

Integrating Powers of x and other Elementary Functions

You can now work out how to integrate any power of x by looking at the corresponding rule for differentiation:

$$\frac{d}{dx}(x^n) = nx^{n-1}, \quad \text{so} \quad \int nx^{n-1}dx = x^n + c.$$

Similarly,

$$\frac{d}{dx}\left(x^{n+1}\right) = (n+1)x^n, \quad \text{so} \quad \int (n+1)x^n dx = x^{n+1} + c.$$

Therefore,

$$\int x^n dx = \int \frac{1}{n+1} \cdot (n+1)x^n dx \qquad \leftarrow \text{notice that } \frac{1}{n+1} \cdot (n+1) \text{ is just 1 when we cancel}$$

$$= \frac{1}{n+1} \int (n+1)x^n dx \qquad \leftarrow \text{take } \frac{1}{n+1} \text{ outside the } \int \text{ sign}$$

$$= \frac{1}{n+1} x^{n+1} + c.$$

You should now look carefully at the formula you have just worked out and ask: for which values of n does it hold? Remember that the differentiation rule $\frac{d}{dx}(x^n) = nx^{n-1}$ holds whether n is positive or negative, a whole number or a fraction or even irrational; in other words, for all real numbers n.

You might expect the integration rule to hold for all real numbers also. But there is one snag: in working it out, we divided by n + 1. Since division by zero does not make sense, the rule will not hold when n + 1 = 0, that is, when n = −1.

Rule 3:

$$\int x^n dx = \frac{1}{n+1} x^{n+1} + c$$

for all real numbers n, except n = −1.

When $n = -1, \int x^n dx$ becomes $\int x^{-1} dx = \int \frac{1}{x} dx$. You don't need to worry that the rule above doesn't apply in this case, because we already know the integral of $\frac{1}{x}$.

Since,

$$\frac{d}{dx}(\ln x) = \frac{1}{x}, \quad \text{we have} \quad \int \frac{1}{x} dx = \ln x + c.$$

Examples: Find,

$$\int x^3 dx$$

$$\int \frac{dx}{x^2}$$

$$\int \sqrt{x} dx$$

Solutions:

$$\int x^3 dx = \frac{1}{(3+1)}x^4 + c = \frac{1}{4}x^4 + c. \leftarrow \text{replacing } n \text{ by } 3 \text{ in the formula.}$$

$$\int \frac{dx}{x^2} = \int x^{-2} dx = \frac{1}{-2+1}x^{-2+1} + c = -\frac{1}{x}c. \qquad \leftarrow \text{replacing } n \text{ by} -2 \text{ in the formula.}$$

$$\int \sqrt{x}\, dx = \int x^{\frac{1}{2}} dx = \frac{1}{\frac{1}{2}+1}x^{\frac{1}{2}+1} + c = \frac{2}{3}x^{\frac{3}{2}} + c. \leftarrow \text{replacing } n \text{ by } \frac{1}{2}.$$

At this stage it is very tempting to give a list of standard integrals, corresponding to the list of derivatives given at the beginning of this booklet. However, you are NOT encouraged to memorise integration formulae, but rather to become VERY familiar with the list of derivatives and to practise recognising a function as the derivative of another function.

If you try memorising both differentiation and integration formulae, you will one day mix them up and use the wrong one. And there is absolutely no need to memorise the integration formulae if you know the differentiation ones.

It is much better to recall the way in which an integral is defined as an anti-derivative. Every time you perform an integration you should pause for a moment and check it by differentiating the answer to see if you get back the function you began with. This is a very important habit to develop. There is no need to write down the checking process every time, often you will do it in your head, but if you get into this habit you will avoid a lot of mistakes.

There is a table of derivatives at the front of this booklet. Try to avoid using it if you can, but refer to it if you are unsure.

Examples:

Find the following indefinite integrals:

$$\int \left(e^x + 3x^{\frac{5}{2}} \right) dx$$

$$\int (5\csc^2 x + 3\sec^2 x)\,dx$$

Solutions:

$$\int \left(e^x + 3x^{\frac{5}{2}} \right) dx = \int e^x dx + 3\int x^{\frac{5}{2}} dx$$

$$= e^x + 3 \cdot \frac{1}{\frac{5}{2}+1} x^{\frac{5}{2}+1} + c$$

$$= e^x + 3 \cdot \frac{2}{7} x^{\frac{7}{2}} + c$$

$$= e^x + \frac{6}{7} x^{\frac{7}{2}} + c.$$

$$\int (5\csc^2 x + 3\sec^2 x)dx = -5\int(-\csc^2 x)dx + 3\int \sec^2 x dx$$

$$= -5\cot x + 3\tan x + c.$$

Using the Chain Rule in Reverse

Recall that, the Chain Rule is used to differentiate composite functions such as $\cos(x^3 + 1)$, $e^{\frac{1}{2}x^2}$, $(2x^2 + 3)^{11}$, $\ln(3x + 1)$. (The Chain Rule is sometimes called the Composite Functions Rule or Function of a Function Rule).

If we observe carefully the answers we obtain when we use the chain rule, we can learn to recognise when a function has this form, and so discover how to integrate such functions.

Remember that, if y = f(u) and u = g(x)

so, $y = f(g(x)),$ (a composite function)

then, $\frac{dy}{dx} = \frac{dy}{du} \cdot \frac{du}{dx}.$

Using function notation, this can be written as,

$$\frac{dy}{dx} = f'(g(x)) \cdot g'(x).$$

In this expression, $f'(g(x))$ is another way of writing $\frac{dy}{du}$ where $y = f(u)$ and $u = g(x)$ and $g'(x)$ is another way of writing $\frac{du}{dx}$ where $u = g(x)$.

This last form is the one you should learn to recognise.

Examples:

By differentiating the following functions, write down the corresponding statement for integration.

$\sin 3x$

$(2x + 1)^7$

ex^2

Solution:

$$\frac{d}{dx}\sin 3x = \cos 3x \cdot 3, \text{ so } \int \cos 3x \cdot 3 dx = \sin 3x + c.$$

$$\frac{d}{dx}(2x+1)^7 = 7(2x+1)^6 \cdot 2, \text{ so } \int 7(2x+1)^6 \cdot 2 dx = (2x+1)^7 + c.$$

$$\frac{d}{dx}\left(e^{x^2}\right) = e^{x^2} \cdot 2x, \text{ so } \int e^{x^2} \cdot 2x dx = e^{x^2} + c.$$

Examples:

Each of the following functions is in the form $f'(g(x)) \cdot g'(x)$.

Identify $f'(u)$ and $u = g(x)$ and hence find an indefinite integral of the function.

$$(3x^2 - 1)^4 \cdot 6x$$

$$\sin(\sqrt{x}) \cdot \frac{1}{2\sqrt{x}}$$

Solutions:

$(3x^2 - 1)^4 \cdot 6x$ is a product of $(3x^2 - 1)^4$ and $6x$.

Clearly $(3x^2 - 1)^4$ is the composite function $f'(g(x))$. So $g(x)$ should be $3x^2 - 1$.

$6x$ is the "other part". This should be the derivative of "what's inside the brackets" i.e. $3x^2 - 1$, and clearly, this is the case:

$$\frac{d}{dx}(3x^2 - 1) = 6x.$$

So, $u = g(x) = 3x^2 - 1$ and $f'(u) = u^4$ giving $f'(g(x)) \cdot g'(x) = (3x^2 - 1)^4 \cdot 6x$. If $f'(u) = u^4$,

$$f(u) = \frac{1}{5}u^5.$$

So, using the rule,

$$\int f'(g(x)) \cdot g'(x) dx = f(g(x)) + c$$

you conclude,

$$\int (3x^3 - 1)^4 \cdot 6x = \frac{1}{5}(3x^2 - 1)^5 + c.$$

You should differentiate this answer immediately and check that you get back the function you began with.

$$\sin(\sqrt{x}) \cdot \frac{1}{2\sqrt{x}}$$

This is a product of $\sin(\sqrt{x})$ and $\frac{1}{2\sqrt{x}}$,

Clearly $\sin(\sqrt{x})$ is a composite function.

The part "inside the brackets" is $\sqrt{x}$ so you would like this to be $g(x)$. The other factor $\dfrac{1}{2\sqrt{x}}$ ought to be $g'(x)$. Let's check if this is the case:

$$g(x) = \sqrt{x} = x^{\frac{1}{2}}, \text{ so } g'(x) = \frac{1}{2}x^{-\frac{1}{2}} = \frac{1}{2x^{\frac{1}{2}}} = \frac{1}{2\sqrt{x}}.$$

Thus, $u = g(x) = \sqrt{x}$ and $f'(u) = \sin u$ giving

$$f'(g(x)) \cdot g'(x) = \sin(\sqrt{x}) \cdot \frac{1}{2\sqrt{x}}.$$

Now, if $f'(u) = \sin u$, $f(u) = -\cos u$.

So, using the rule

$$\int f'(g(x)) \cdot g'(x)dx = f(g(x)) + c$$

you conclude,

$$\int \sin(\sqrt{x}) \cdot \frac{1}{2\sqrt{x}}dx = -\cos(\sqrt{x}) + c.$$

Example:

Integrate $\int \sin^3 x \cdot \cos x dx$.

Solution:

$$\int \sin^3 x \cdot \cos x dx = \int (\sin x)^3 \cdot \cos x dx.$$

So $u = g(x) = \sin x$ with $g'(x) = \cos x$.

And $f'(u) = u^3$ giving $f(u) = \frac{1}{4}u^4$.

Hence $\int \sin^3 x \cdot \cos x dx = \frac{1}{4}(\sin x)^4 + c = \frac{1}{4}\sin^4 x + c.$

Applications

Applications of anti-differentiation arise in problems in which you know the rate of change of a function and want to find the function itself. Problems about motion provide many examples, such as those in which the velocity of a moving object is given and you want to find its position at any time. Since velocity is rate of change of displacement, you must anti-differentiate to find the displacement.

Examples:

1. A stone is thrown upwards from the top of a tower 50 metres high. Its velocity in an upwards direction t second later is 20 – 5t metres per second. Find the height of the stone above the ground 3 seconds later.

Solution: Let the height of the stone above ground level at time t be h metres.

We are given two pieces of information in this problem:

- The fact that the tower is 50m high tells us that when t = 0 (that is, at the moment the stone leaves the thrower's hand), h = 50,

- The fact that the velocity at time t is 20 − 5t tells us that

$$\frac{dh}{dt} = 20 - 5t.$$

We begin with the second statement, which tells us about rate of change of a function. By anti-differentiating, we obtain

$$h = \int (20 - 5t) dt$$
$$= 20t - \frac{5}{2}t^2 + c.$$

It is vitally important not to forget the '+c' (the constant of integration) in problems like this.

Now we can make use of the first statement, which is called an initial condition (it tells us what things were like at the start) to find a value for the constant c. By substituting h = 50 and t = 0 into the equation

$$h = \int (20 - 5t) dt$$
$$= 20t - \frac{5}{2}t^2 + c,$$

you obtain

$$50 = c.$$

$$\text{So } h = 20t - \frac{5}{2}t^2 + 50.$$

Finally, let us go back to the problem, and read it again, to check exactly what we are asked to find: 'Find the height of the stone above the ground 3 seconds later'. To find this, all we have to do is substitute t = 3 in the expression we have just derived.

When $t = 3, h = 60 - \frac{5}{2} \cdot 9 + 50 = 87.5$, the height of the stone 3 seconds later is 87.5 metres.

Let us look back at the structure of this problem and its solution:

- We are given information about the rate of change of a quantity, and we antidifferentiate (i.e. integrate) to get a general expression for the quantity, including an arbitary constant.

- We are given information about initial conditions and use this to find the value of the constant of integration.

- We now have a precise expression for the quantity, and can use that to answer the questions asked.

2. When a tap at the base of a storage tank is turned on, water flows out of the tank at the rate of $200e^{-\frac{1}{5}t}$ litres per minute. If the volume of water in the tank at the start is 1000 litres, find how much is left after the tap has been running for 10 minutes.

Solution:

Let the volume of water in the tank t minutes after the tap is turned on be V litres. Since water is running out of the tank, V will be decreasing, and so $\dfrac{dV}{dt}$ must be negative.

Hence,

$$\frac{dV}{dt} = -200e^{-\frac{1}{5}t}$$

and so,

$$V = \int \left(-200e^{-\frac{1}{5}t} \right) dt$$

$$= -200 \cdot (-5) \int e^{-\frac{1}{5}t} \left(-\frac{1}{5} \right) dt$$

$$= 1000e^{-\frac{1}{5}t} + c.$$

Now, when $t = 0$, $V = 1000$, so $1000 = 1000 + c$, hence $c = 0$.

So, $V = 1000e^{-\frac{1}{5}t}$

Thus, when $t = 10$, $V = 1000e^{-2}$

$$\approx 135.34 \, litres.$$

Differential Equation

A differential equation is any equation which contains derivatives, either ordinary derivatives or partial derivatives.

There is one differential equation that everybody probably knows, that is Newton's Second Law of Motion. If an object of mass m is moving with acceleration a and being acted on with force F then Newton's Second Law tells us:

F = ma.

To see that this is in fact a differential equation you need to rewrite it a little. First, remember that we can rewrite the acceleration, a, in one of two ways.

$$a = \frac{dv}{dt} \;\; \text{OR} \;\; a = \frac{d^2u}{dt^2}$$

Where v is the velocity of the object and u is the position function of the object at any time t. You should also remember at this point that the force, F may also be a function of time, velocity, and position.

So, with all these things in mind Newton's Second Law can now be written as a differential equation in terms of either the velocity, v, or the position, u, of the object as follows:

$$m\frac{dv}{dt} = F(t,v)$$

$$m\frac{d^2u}{dt^2} = F\left(t, u, \frac{du}{dt}\right).$$

So, here is our first differential equation. Here are a few more examples of differential equations.

$$ay'' + by' + cy = g(t),$$

$$\sin(y)\frac{d^2y}{dx^2} = (1-y)\frac{dy}{dx} + y^2 e^{-5y},$$

$$y^{(4)} + 10y''' - 4y' + 2y = \cos(t),$$

$$\alpha^2 \frac{\partial^2 u}{\partial x^2} = \frac{\partial u}{\partial t},$$

$$a^2 u_{xx} = u_{tt},$$

$$\frac{\partial^3 u}{\partial^2 x \partial t} = 1 + \frac{\partial u}{\partial y}.$$

Ordinary and Partial Differential Equations

A differential equation is called an ordinary differential equation, abbreviated by ode, if it has ordinary derivatives in it. Likewise, a differential equation is called a partial differential equation, abbreviated by pde, if it has partial derivatives in it.

Linear Differential Equations

A linear differential equation is any differential equation that can be written in the following form:

$$a_n(t)y^{(n)}(t) + a_{n-1}(t)y^{(n-1)}(t) + \cdots + a_1(t)y'(t) + a_0(t)y(t) = g(t).$$

The important thing to note about linear differential equations is that there are no products of the function, $y(t)$, and its derivatives and neither the function or its derivatives occur to any power other than the first power. Also note that neither the function or its derivatives are "inside" another function, for example, $\sqrt{y'}$ or e^y. The coefficients $a_0(t),...,a_n(t)$ and $g(t)$ can be zero or non-zero functions, constant or non-costant functions, linear or non-linear functions. Only the function $y(t)$, and its derivatives are used in determining if a differential equation is linear.

If a differential equation cannot be written in the form of equation $a_n(t)y^{(n)}(t)+a_{n-1}(t)y^{(n-1)}(t)+...+a_1(t)y'(t)+a_0(t)y(t)=g(t)$ then it is called a non-linear differential equation.

$\sin(y)\dfrac{d^2y}{dx^2}=(1-y)\dfrac{dy}{dx}+y^2e^{-5y}$ is non-linear, the other two are linear differential equations. We can't classify ($m\dfrac{dv}{dt}=F(t,v)$) and ($m\dfrac{d^2u}{dt^2}=F\left(t,u,\dfrac{du}{dt}\right)$) since we do not know what form the function F has. These could be either linear or non-linear depending on F.

Solution:

A solution to a differential equation on an interval $\alpha < t < \beta$ is any function $y(t)$ which satisfies the differential equation in question on the interval $\alpha < \ < \beta$. It is important to note that solutions are often accompanied by intervals and these intervals can impart some important information about the solution. Consider the following example.

Example: Show that $y(x)=x^{-\frac{3}{2}}$ is a solution to $4x^2y''+12xy'+3y=0$ for >0.

Solution:

You'll need the first and second derivative to do this.

$$y'(x)=-\frac{3}{2}x^{-\frac{5}{2}} \quad y''(x)=\frac{15}{4}x^{-\frac{7}{2}}.$$

Plug these as well as the function into the differential equation.

$$4x^2\left(\frac{15}{4}x^{-\frac{7}{2}}\right)+12x\left(-\frac{3}{2}x^{-\frac{5}{2}}\right)+3\left(x^{-\frac{3}{2}}\right)=0$$

$$15x^{-\frac{3}{2}}-18x^{-\frac{3}{2}}+3x^{-\frac{3}{2}}=0$$

$$0=0.$$

So, $y(x)=x^{-\frac{3}{2}}$ does satisfy the differential equation and hence is a solution. Why then did you include the condition that $x>0$? You did not use this condition anywhere in the work showing that the function would satisfy the differential equation.

To see why recall that,

$$y(x) = x^{-\frac{3}{2}} \frac{1}{\sqrt{x^3}}$$

In this form it is clear that you'll need to avoid $x = 0$ at the least as this would give division by zero.

Also, there is a general rule of thumb that we're going to run with in this class. This rule of thumb is: Start with real numbers, end with real numbers. In other words, if our differential equation only contains real numbers then we don't want solutions that give complex numbers. So, in order to avoid complex numbers you will also need to avoid negative values of x.

So, you saw in the last example that even though a function may symbolically satisfy a differential equation, because of certain restrictions brought about by the solution we cannot use all values of the independent variable and hence, must make a restriction on the independent variable. This will be the case with many solutions to differential equations.

In the last example, note that there are in fact many more possible solutions to the differential equation given. For instance, all of the following are also solutions:

$$y(x) = x^{-\frac{1}{2}}.$$

$$y(x) = -9x^{-\frac{3}{2}}.$$

$$y(x) = 7x^{-\frac{1}{2}}.$$

$$y(x) = -9x^{-\frac{3}{2}} + 7x^{-\frac{1}{2}}.$$

Given these examples can you come up with any other solutions to the differential equation? There are in fact an infinite number of solutions to this differential equation.

Initial Conditions

Initial Conditions are a condition, or set of conditions, on the solution that will allow us to determine which solution that you are after. Initial conditions (often abbreviated i.c.'s when you're feeling lazy) are of the form,

$$y(t_0) = y_0 \quad \text{and/ or } y^{(k)}(t_0) = y_k.$$

So, in other words, initial conditions are values of the solution and its derivatives at specific points. As you will see eventually, solutions to differential equations are unique and hence only one solution will meet the given initial conditions.

The number of initial conditions that are required for a given differential equation will depend upon the order of the differential equation.

Example: $y(x) = x^{-\frac{3}{2}}$ is a solution to $4x^2 y'' + 12xy' + 3y = 0$, $y(4) = \frac{1}{8}$, and $y'(4) = -\frac{3}{64}$.

Solution: As you saw in previous example the function is a solution and you can then note that,

$$y(4) = 4^{-\frac{3}{2}} = \frac{1}{\left(\sqrt{4}\right)^3} = \frac{1}{8}$$

$$y'(4) = -\frac{3}{2}4^{-\frac{5}{2}} = -\frac{3}{2}\frac{1}{\left(\sqrt{4}\right)^5} = -\frac{3}{64}$$

So, this solution also meets the initial conditions of $y(4) = \frac{1}{8}$ and $y'(4) = -\frac{3}{64}$. In fact, $y(x) = x^{-\frac{3}{2}}$ is the only solution to this differential equation that satisfies these two initial conditions.

Initial Value Problem

An Initial Value Problem (or IVP) is a differential equation along with an appropriate number of initial conditions. The number of initial conditions.

As we noted earlier the number of initial conditions required will depend on the order of the differential equation.

Interval of Validity

The interval of validity for an IVP with initial conditions,

$$y(t_0) = y_0 \quad \text{and} \quad y^{(k)}(t_0) = y_k,$$

is the largest possible interval on which the solution is valid and contains toto. These are easy to define, but can be difficult to find, so you're going to put off saying anything more about these until you get into actually solving differential equations and need the interval of validity.

General Solution

The general solution to a differential equation is the most general form that the solution can take and doesn't take any initial conditions into account.

Actual Solution

The actual solution to a differential equation is the specific solution that not only satisfies the differential equation, but also satisfies the given initial conditions.

Example: What is the actual solution to the following IVP?

$$2t\,y' + 4y = 3 \quad y(1) = -4.$$

Solution: This is actually easier to do than it might at first appear. From the previous example you already know that all solutions to the differential equation are of the form.

$$y(t) = \frac{3}{4} + \frac{c}{t^2}$$

All that you need to do is determine the value of cc that will give us the solution that you're after. To find this all you need do is use our initial condition as follows:

$$-4 = y(1) = \frac{3}{4} + \frac{c}{1^2} \Rightarrow c = -4 - \frac{3}{4} = -\frac{19}{4}$$

So, the actual solution to the IVP is:

$$y(t) = \frac{3}{4} - \frac{19}{4t^2}.$$

From this last example you can see that once you have the general solution to a differential equation finding the actual solution is nothing more than applying the initial conditions and solving for the constants that are in the general solution.

Implicit/Explicit Solution

In this case it's easier to define an explicit solution, then tell you what an implicit solution isn't and then give you an example to show you the difference. So, that's what you'll do.

An explicit solution is any solution that is given in the form $y = y(t)$. In other words, the only place that y actually shows up is once on the left side and only raised to the first power. An implicit solution is any solution that isn't in explicit form. It is possible to have either general implicit/explicit solutions and actual implicit/explicit solutions.

Example: Find an actual explicit solution to $y' = \frac{1}{y}$, $y(2) = -1$.

Solution: You already know from the previous example that an implicit solution to this IVP is $y^2 = t^2 - 3$. To find the explicit solution all you need to do is solve for $y(t)$.

$$y(t) = \pm\sqrt{t^2 - 3}$$

Now, you've got a problem here. There are two functions here and you only want one and in fact only one will be correct. You can determine the correct function by reapplying the initial condition. Only one of them will satisfy the initial condition.

In this case you can see that the solution will be the correct one. The actual explicit solution is then,

$$y(t) = -\sqrt{t^2 - 3}.$$

In this case you were able to find an explicit solution to the differential equation. It should be noted however that it will not always be possible to find an explicit solution.

Also, note that in this case you were only able to get the explicit actual solution because you had the initial condition to help us determine which of the two functions would be the correct solution.

Order and Degree of Differential Equation

Differential Equations are classified on the basis of the order. Order of a differential equation is the order of the highest order derivative (also known as differential coefficient) present in the equation.

For Example:

$$\frac{d^3x}{dx^3} + 3x\frac{dy}{dx} = e^y.$$

In this equation the order of the highest derivative is 3 hence this is a third order differential equation.

Example : $-\left(\frac{d^2y}{dx^2}\right)^4 + \frac{dy}{dx} = 3.$

This equation represents a fourth order differential equation.

This way we can have higher order differential equations i.e. n^{th} Order differential equations.

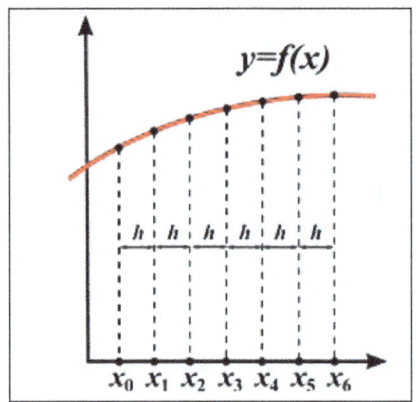

First Order Differential Equation

The order of highest derivative in case of first order differential equations is 1. A linear differential equation has order 1. In case of linear differential equations, the first derivative is the highest order derivative.

$$\frac{dy}{dx} + Py = Q.$$

P and Q are either constants or functions of the independent variable only.

This represents a linear differential equation whose order is 1.

Example : $\frac{dy}{dx} + (x^2 + 5)y = \frac{x}{5}$

This also represents a First order Differential Equation.

Second Order Differential Equation

When the order of the highest derivative present is 2, then it represents a second order differential equation.

Example: $\dfrac{d^2y}{dx^2} + (x^3 + 3x)y = 9$.

In this example, the order of the highest derivative is 2. Therefore, it is a second order differential equation.

Degree of Differential Equation

The degree of differential equation is represented by the power of the highest order derivative in the given differential equation.

The differential equation must be a polynomial equation in derivatives for the degree to be defined.

Example : $\dfrac{d^4y}{dx^4} + \left(\dfrac{d^2y}{dx^2}\right)^2 - 3\dfrac{dy}{dx} + y = 9$.

Here, the exponent of the highest order derivative is one and the given differential equation is a polynomial equation in derivatives. Hence, the degree of this equation is 1.

Example: $[\dfrac{d^2y}{dx^2} + (\dfrac{dy}{dx})^2]^4 = k^2(\dfrac{d^3y}{dx^3})^2$.

The order of this equation is 3 and the degree is 2 as the highest derivative is of order 3 and the exponent raised to highest derivative is 2.

Example: $\dfrac{d^2y}{dx^2} + \cos\dfrac{d^2y}{dx^2} = 5x$.

The given differential equation is not a polynomial equation in derivatives. Hence, the degree for this equation is not defined.

Example: $\left(\dfrac{d^3y}{dx^3}\right)^2 + y = 0$.

The order of this equation is 3 and the degree is 2.

Example: Figure out the order and degree of differential equation that can be formed from the equation $\sqrt{1-x^2} + \sqrt{1-y^2} = k(x-y)$.

Solution: Let $x = \sin\theta$, $y = \sin\phi$.

So, the given equation can be rewritten as,

$$\sqrt{1-\sin\theta^2} + \sqrt{1-\sin\phi^2} = k(\sin\theta - \sin\phi)$$

$$\Rightarrow (\cos\theta + \cos\phi) = k(\sin\theta - \sin\phi)$$

$$\Rightarrow 2\cos\frac{\theta+\phi}{2}\cos\frac{\theta-\phi}{2} = 2k\cos\frac{\theta+\phi}{2}\sin\frac{\theta-\phi}{2}$$

$$\cot\frac{\theta-\phi}{2} = k$$

$$\theta - \phi = 2\cot^{-1}k$$

$$\sin^{-1}x - \sin^{-1}y = 2\cot^{-1}k.$$

Differentiating both sides w. r. t. x, you get,

$$\frac{1}{1-x^2} - \frac{1}{1-y^2} = \frac{dy}{dx}.$$

So, the degree of the differential equation is 1 and it is a first order differential equation.

References

- Differentiation-mathematics: britannica.com, Retrieved 20 January, 2019

- DifferentiationRevision, mathslearning: adelaide.edu.au, Retrieved 28 July, 2019

- Partial-Derivatives, Calculus: maths.tcd.ie, Retrieved 19 March, 2019

- Integration-mathematics: britannica.com, Retrieved 02 Februeary, 2019

- Integration, maths-learning-centre: sydney.edu.au, Retrieved 16 April, 2019

- Differential-equation-and-its-degree: byjus.com, Retrieved 16 March, 2019

Ordinary Differential Equations of First Order and First Degree

Ordinary differential equations of first order include a(x) and f(x) as the continuous functions of x. The two methods of solving such differential equations are integrating factor and method of variation of a constant. This chapter discusses in detail about the ordinary differential equations of first order and first degree.

Ordinary Differential Equations

An ordinary differential equation (ODE) is an equation that involves some ordinary derivatives (as opposed to partial derivatives) of a function. Often, our goal is to *solve* an ODE, i.e., determine what function or functions satisfy the equation.

If you know what the derivative of a function is, how can you find the function itself? You need to find the antiderivative, i.e., you need to integrate. For example, if you are given:

$$\frac{dx}{dt}(t) = \cos t$$

then what is the function $x(t)$? Since the antiderivative of $\cos t$ is $\sin t$, then $x(t)$ must be $\sin t$. Except we forgot one important point: there is always an arbitrary constant that we cannot determine if we only know the derivative. Therefore, all we can detetermine from the above equation is that:

$$x(t) = \sin t + C$$

for some arbitrary constant C. You can verify that indeed $x(t)$ satisfies the equation $\frac{dx}{dt} = \cos t$.

In general, solving an ODE is more complicated than simple integration. Even so, the basic principle is always integration, as we need to go from derivative to function. Usually, the difficult part is determining what integration we need to do.

The Simplest Possible ODE

Let's start simpler, though. What is the simplest possible ODE? Let $x(t)$ be a function of t that satisfies the ODE:

$$\frac{dx}{dt} = 0.$$

We can ask some simple questions. What is $x(t)$? is $x(t)$ uniquely determined from this equation? If not, what else do you need to specify?

Equation $\frac{dx}{dt} = 0$ just means that $x(t)$ is a constant function, $x(t) = C$. It is certainly not uniquely determined, as there is no way to specify the constant C if we only have equations for the erivatives of x. In order to uniquely determine $x(t)$, one must provide some additional data in terms of the function $x(t)$ itself.

We could for example, specify that $x(t)$ must be equal to 31 when $t = 11$, adding the condition:

$$x(11) = 31.$$

Then we know $C = 31$ and the function is $x(t) = 31$ as an initial condition.

Let's write the initial condition more generally as:

$$x(t_0) = x_0,$$

Where t_0 is some given time and x_0 is some given number. It's as though we initialize the system to be equal to the number x_0 at the time $t = t_0$. However, this "initial condition" also determines $x(t)$ for early times. As you can see from the solution $x(t) = 31$ for all time t, this condition specifies the state of the system for times before and after $t = 11$.

A Slightly more Complicated ODE

Let's make things a little more complicated. Consider the equation:

$$\frac{dx}{dt} = m \sin t + nt^3,$$

Where m and n are just some real numbers. Equation $\frac{dx}{dt} = m \sin t + nt^3$, above isn't much more complicated than equation $\frac{dx}{dt} = 0$ because the right hand side does not depend on x. It only depends on t. We are simply specifying what the derivative is in terms of t. The solution is just the antiderivative, or the integral.

Let's do the integral slightly differently this time. We'll use the definite integral from time $t = a$ to time $t = b$. Using the fundamental theorem of calculus, the integral of $\frac{dx}{dt}$ from a to b must be:

$$x(b) - x(a) = \int_a^b \frac{dx}{dt} dt$$
$$\int_a^b (m \sin t + nt^3) dt$$
$$= -m \cos b + nb^4 / 4 - (-m \cos a + na^4 / 4).$$

We can write the solution in different ways. We could just replace b with an arbitrary time t,

$$x(t) = -m \cos t + nt^4 / 4 + m \cos a - na^4 / 4 + x(a).$$

This form makes it very obvious how the solution $x(t)$ would depend on an initial condition $x(t_0) = x_0$. If $x(7) = 5$, then:

$$x(t) = -m\cos t + nt^4/4 + m\cos 7 - n7^4/4 + 5.$$

On the other hand, if we aren't concerned with the form of the constant, we could just write the general solution as:

$$x(t) = -m\cos t + nt^4/4 + C$$

for some arbitrary constant C.

An ODE that isn't a Simple Integral

So far, the example ODEs we've seen could be solved simply by integrating. The reason they were so simple is that the equations for $\frac{dx}{dt}$ did not depend on the function $x(t)$ but only on the variable t. On the other hand, once the equation depends on both $\frac{dx}{dt}$ and $x(t)$, we have do more work to solve for the function $x(t)$.

Here's an ODE that includes $x(t)$:

$$\frac{dx}{dt} = ax(t) + b$$

Where a and b are some constants. Since the right hand side depends on x itself, we cannot simply integrate and use the fundamental theorem of calculus. To solve this ODE for $x(t)$ we'll need to do some manipulations and use the chain rule (i.e., a u-substitution).

The first thing to do is get all expressions involving x on one side of the equation. If we subtract, we won't be able to put things in the right form for the chain rule, as we'll have terms without a $\frac{dx}{dt}$ in them. Instead, we divide both sides of the equation by $ax(t) + b$,

$$\frac{\frac{dx}{dt}}{ax(t) + b} = 1.$$

Now the right hand side is a simple function of t (a constant function in this case). We can integrate both sides of the equation with respect to t,

$$\int \frac{\frac{dx}{dt}\,dt}{ax(t) + b} = \int 1\,dt.$$

It is in a special form that makes it easy to integrate. It contains a $\frac{dx}{dt}\,dt$ factor, and the remaining dependence on t is only through the function $x(t)$. If we change variables (do a u-substitution) of the form $u = x(t)$, then $du = \dfrac{dx}{dt}dt$, and we just replace the remaining appearances of $x(t)$ with u.

The left hand side is then a simple integral in terms of the new variable u, which we can integrate and substitute back $u = x(t)$:

$$\int \frac{\dfrac{dx}{dt}dt}{ax(t)+b} = \int \frac{du}{au+b}$$

$$= \frac{1}{a}\log|au+b| + C_1$$

$$= \frac{1}{a}\log|ax(t)+b| + C_1,$$

for some arbitrary constant C_1.

Since this expression must be equal to $\int 1 dt = t + C_2$ for another arbitrary constant C_2 we obtain an equation for $x(t)$ and t,

$$\frac{1}{a}\log|ax(t)+b| + C_1 = t + C_2.$$

Let $C_3 + C_2 - C_1$ and then solve the equation for $x(t)$:

$$\frac{1}{a}\log|ax(t)+b| = t + C_3$$

$$|ax(t)+b| = \exp(at + aC_3)$$

$$|ax(t)+b| = \pm\exp(at + aC_3)$$

$$x(t) = \pm\frac{1}{a}\exp(at + aC_3) - b/a.$$

(The notation exp (z) is just another way of writing the exponential e^z). We can write this equation more simply by defining a new arbitrary constant $C = \pm\dfrac{1}{a}\exp(aC_3)$. Then, the solution to our ODE can be written as:

$$x(t) = Ce^{at} - b/a.$$

Can you verify that this solution for $x(t)$ does indeed satisfy the original ODE $\dfrac{dx}{dt} = ax + b$? Since checking that a solution satisfies an ODE is much easier and less error-prone than solving the ODE, verifying the solution is an essential step in the solution process.

If $x = Ce^{at} - b/a$, then $\dfrac{dx}{dt} = Cae^{at}$. On the other hand, $ax + b = Cae^{at} - b + b = Cae^{at}$. Yes, these expressions match, $\dfrac{dx}{dt} = ax + b$, and we can be confident of our solution.

In order to determine the constant C, we need an additional condition. For example, if $x(3) = 4$, then C must satisfy:

$$4 = Ce^{3a} - b/a$$

So that,

$$C = (4 + b/a)e^{-3a}.$$

Our solution for this initial condition is:

$$x(t) = (4 + b/a)e^{-3a}e^{at} - b/a$$

or

$$x(t) = (4 + b/a)e^{a(t-3)} - b/a.$$

A Shortcut Method to Solving Simple ODES

For the above solution, we did some extra steps in order to demonstrate that the manipulations were really nothing more than a u-substitution. Usually, we'll skip many of these steps and use a shortcut method. However, before jumping into the shortcut method, make sure you understand how the above u-substitution works.

Let's revisit our solution method to see how we can take some shortcuts. The first thing we could do differently is avoid changing to the variable u. We could keep everything in terms of x, in which case, the u-substitution would be replacing $x(t)$ with x and $\frac{dx}{dt}dt$ with dx.

Next, observe the results of the substitution. We started with,

$$\frac{\frac{dx}{dt}}{ax+b} = 1$$

and ended up with,

$$\int \frac{dx}{ax+b} = \int 1 dt,$$

where now we wrote everything in terms of x rather than u. To accomplish this manipulation, we multiplied by dt and did our substitution to replace $\frac{dx}{dt}dt$ by dx. It was as though we canceled the dt from the numerator with the dt from the denominator. The derivative $\frac{dx}{dt}$ isn't really a fraction of numbers dx and dt but in an integral, applying the chain rule (i.e., u-substitution) makes it behave like it is a fraction.

Hence, in practice, we can safely treat $\frac{dx}{dt}$ like a fraction when used in this context of forming an integral to solve a differential equation. To solve the equation $\dfrac{dx}{dt} = ax + b,$ we multiply both sides of the equation by dt and divide both sides of the equation by $ax + b$ to get:

$$\frac{dx}{ax+b} = dt.$$

Then, we integrate both sides to obtain:

$$\int \frac{dx}{ax+b} = \int dt.$$

Just remember that these manipulations are really a shortcut way to denote using the chain rule.

The simple ODEs of this introduction give you a taste of what ordinary differential equations are and how we can solve them.

Differential Equations of First Order and First Degree

Any ordinary differential equation of the first order and first degree can be written in the form:

$$\frac{dy}{dx} = F(x, y) \;\; or \;\; M(x, y)dx + N(x, y)dy = 0$$

Example: The differential equation,

$$\frac{dy}{dx} = \frac{x - 3y}{2y - x}$$

can also be written as,

$$(x - 3y)dx + (x - 2y)dy = 0$$

Existence of a Solution

The general solution of the equation dy/dx = g(x, y), if it exists, has the form f(x, y, C) = 0, where C is an arbitrary constant. Under what circumstances does a general solution exist? We have the following theorem.

Theorem 1: A general solution of dy/dx = g(x, y) exists over some specified region R of points (x, y) if the following conditions are met:

- g(x, y) is continuous and single-valued over R

- ∂g/∂y exists and is continuous at all points of R

The general solution f(x, y, C) = 0 of a differential equation dy/dx = g(x, y) over some region R consists of a family of curves, called the integral curves of the differential equation, (one curve for each possible value of C, each curve representing a particular solution), such that through each point in R there passes one and only one curve of the family f(x, y, C) = 0.

The differential equation:

$$\frac{dy}{dx} = g(x, y)$$

associates with each point (x_0, y_0) in the region R a direction:

$$m = \frac{dy}{dx}\bigg|_{x_0, y_0} = g(x_0, y_0)$$

The direction at each point of R is that of the tangent to that curve of the family f(x, y, C) = 0 that passes through the point.

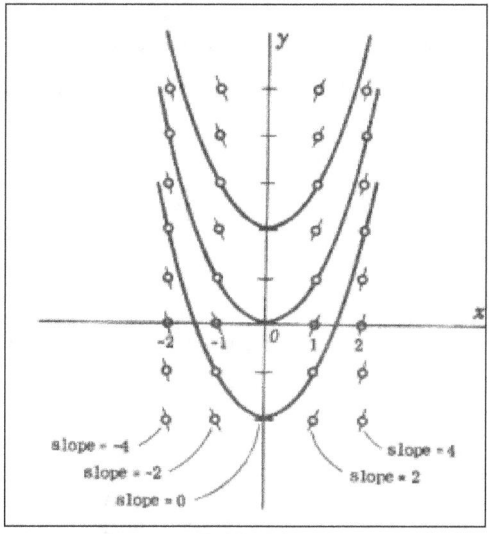

Direction field of dy/dx = 2x.

A region R in which a direction is associated with each point is called a direction field. In figure is shown the direction field and integral curves for the differential equation dy/dx = 2x. The general solution of this equation is $y = x^2 + C$. The integral curves are parabolas.

Separable Equations

A separable differential equation is any differential equation that we can write in the following form.

$$N(y)\frac{dy}{dx} = M(x)$$

In order for a differential equation to be separable all the y's in the differential equation must be multiplied by the derivative and all the x's in the differential equation must be on the other side of the equal sign.

To solve this differential equation we first integrate both sides with respect to x to get,

$$\int N(y)\frac{dy}{dx}dx = \int M(x)dx$$

Now, remember that y is really y(x) and so we can use the following substitution,

$$u = y(x) \quad du = y'(x)\,dx = \frac{dy}{dx}\,dx$$

Applying this substitution to the integral we get,

$$\int N(u)du = \int M(x)dx$$

At this point we can integrate both sides and then back substitute for the u on the left side. As implied in the previous sentence, it might not actually be possible to evaluate one or both of the integrals at this point. If that is the case, then there won't be a lot we can do to proceed using this method to solve the differential equation.

Now, the process above is the mathematically correct way of doing solving this differential equation. Note however, that if we "separate" the derivative as well we can write the differential equation as,

$$N(y)dy = M(x)dx$$

We obviously can't separate the derivative like that, but let's pretend we can for a bit and we'll see that we arrive at the answer with less work.

Now we integrate both sides of this to get,

$$\int N(y)dy = \int M(x)dx$$

So, if we compare $\int N(u)du = \int M(x)dx$ and $\int N(y)dy = \int M(x)dx$ we can see that the only difference is on the left side and even then the only real difference is that equation $\int N(u)du = \int M(x)dx$ has the integral in terms of u and equation $\int N(y)dy = \int M(x)dx$ has the integral in terms of y. Outside of that there is no real difference. The integral on the left is exactly the same integral in each equation. The only difference is the letter used in the integral. If we integrate equation $\int N(u)du = \int M(x)dx$ and then back substitute in for u we would arrive at the same thing as if we'd just integrated equation $\int N(y)dy = \int M(x)dx$ from the start.

Therefore, to make the work go a little easier, we'll just use equation $\int N(y)dy = \int M(x)dx$ to find the solution to the differential equation. Also, after doing the integrations, we will have an implicit solution that we can solve for the explicit solution, y(x). Note that it won't always be possible to solve for an explicit solution.

An implicit solution is a solution that is not in the form y=y(x) while an explicit solution has been written in that form.

We will also have to worry about the interval of validity for many of these solutions. Recall that the interval of validity was the range of the independent variable, x in this case, on which the solution is valid. In other words, we need to avoid division by zero, complex numbers, logarithms of negative numbers or zero, *etc.* Most of the solutions that we will get from separable differential equations will not be valid for all values of x.

Let's start things off with a fairly simple example so we can see the process without getting lost in details of the other issues that often arise with these problems.

Example: Solve the following differential equation and determine the interval of validity for the solution.

$$\frac{dy}{dx} = 6y^2 x \quad y(1) = \frac{1}{25}$$

Solution: It is clear that this differential equation is separable. So, let's separate the differential equation and integrate both sides. As with the linear first order officially we will pick up a constant of integration on both sides from the integrals on each side of the equal sign. The two can be moved to the same side and absorbed into each other. We will use the convention that puts the single constant on the side with the x's given that we will eventually be solving for y and so the constant would end up on that side anyway.

$$y^{-2} dy = 6x \, dx$$
$$\int y^{-2} dy = \int 6x \, dx$$
$$-\frac{1}{y} = 3x^2 + c$$

So, we now have an implicit solution. This solution is easy enough to get an explicit solution, however before getting that it is usually easier to find the value of the constant at this point. So apply the initial condition and find the value of c.

$$-\frac{1}{\frac{1}{25}} = 3(1)^2 + c \quad c = -28$$

Plug this into the general solution and then solve to get an explicit solution.

$$-\frac{1}{y} = 3x^2 - 28$$
$$y(x) = \frac{1}{28 - 3x^2}$$

Now, as far as solutions go we've got the solution. We do need to start worrying about intervals of validity however.

Recall that there are two conditions that define an interval of validity. First, it must be a continuous interval with no breaks or holes in it. Second it must contain the value of the independent variable in the initial condition, $x = 1$ in this case.

So, for our case we've got to avoid two values of x. Namely, $x \neq \pm\sqrt{\dfrac{28}{3}} \approx \pm 3.05505$ since these will give us division by zero. This gives us three possible intervals of validity.

$$-\infty < x < -\sqrt{\frac{28}{3}} \qquad -\sqrt{\frac{28}{3}} < x < \sqrt{\frac{28}{3}} \qquad \sqrt{\frac{28}{3}} < x < \infty$$

However, only one of these will contain the value of xx from the initial condition and so we can see that:

$$-\sqrt{\frac{28}{3}} < x < \sqrt{\frac{28}{3}}$$

must be the interval of validity for this solution.

Here is a graph of the solution.

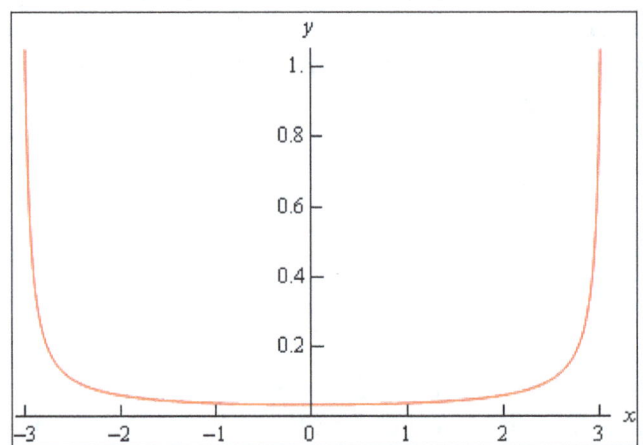

This does not say that either of the other two intervals can't be the interval of validity for any solution to the differential equation. With the proper initial condition either of these could have been the interval of validity.

We'll leave it to you to verify the details of the following claims. If we use an initial condition of:

$$y(-4) = -\frac{1}{20}$$

we will get exactly the same solution however in this case the interval of validity would be the first one.

$$-\infty < x < -\sqrt{\frac{28}{3}}$$

Likewise, if we use:

$$y(6) = -\frac{1}{80}$$

as the initial condition we again get exactly the same solution and, in this case, the third interval becomes the interval of validity.

$$\sqrt{\frac{28}{3}} < x < \infty$$

So, simply changing the initial condition a little can give any of the possible intervals.

Example: Solve the following IVP and find the interval of validity for the solution.

$$y' = \frac{3x^2 + 4x - 4}{2y - 4} \quad y(1) = 3$$

Solution: This differential equation is clearly separable, so let's put it in the proper form and then integrate both sides.

$$(2y - 4)dy = (3x^2 + 4x - 4)dx$$
$$\int (2y - 4)dy = \int (3x^2 + 4x - 4)dx$$
$$y^2 - 4y = x^3 + 2x^2 - 4x + c$$

We now have our implicit solution, so as with the first example let's apply the initial condition at this point to determine the value of c.

$$(3)^2 - 4(3) = (1)^3 + 2(1)^2 - 4(1) + c \quad c = -2$$

The implicit solution is then:

$$y^2 - 4y - (x^3 + 2x^2 - 4x - 2) = 0$$

We now need to find the explicit solution. This is actually easier than it might look and you already know how to do it. First, we need to rewrite the solution a little:

$$y^2 - 4y - (x^3 + 2x^2 - 4x - 2) = 0$$

To solve this all we need to recognize is that this is quadratic in y and so we can use the quadratic formula to solve it. However, unlike quadratics you are used to, at least some of the "constants" will not actually be constant but will in fact involve x's.

So, upon using the quadratic formula on this we get.

$$y(x) = \frac{4 \pm \sqrt{16 - 4(1)(-(x^3 + 2x^2 - 4x - 2))}}{2}$$
$$= \frac{4 \pm \sqrt{16 + 4(x^3 + 2x^2 - 4x - 2)}}{2}$$

Next, We can factor a 4 out from under the square root (it will come out as a 2) and then simplify a little.

$$y(x) = \frac{4 \pm 2\sqrt{4 + (x^3 + 2x^2 - 4x - 2)}}{2}$$
$$= 2 \pm \sqrt{x^3 + 2x^2 - 4x + 2}$$

We are almost there. We've actually got two solutions here (the "±") and we only want a single solution. In fact, only one of the signs can be correct. So, to figure out which one is correct we can reapply the initial condition to this. Only one of the signs will give the correct value so we can use this to figure out which one of the signs is correct. Plugging x = 1 into the solution gives.

$$3 = y(1) = 2 \pm \sqrt{1 + 2 - 4 + 2} = 2 \pm 1 = 3, 1$$

In this case it looks like the "+" is the correct sign for our solution. Note that it is completely possible that the "−" could be the solution (*i.e.* using an initial condition of y(1)=1) so don't always expect it to be one or the other.

The explicit solution for our differential equation is:

$$y(x) = 2 + \sqrt{x^3 + 2x^2 - 4x + 2}$$

To finish the example out we need to determine the interval of validity for the solution. If we were to put a large negative value of x in the solution we would end up with complex values in our solution and we want to avoid complex numbers in our solutions here. So, we will need to determine which values of x will give real solutions. To do this we will need to solve the following inequality.

$$x^3 + 2x^2 - 4x + 2 \geq 0$$

In other words, we need to make sure that the quantity under the radical stays positive.

Using a computer algebra system like Maple or Mathematical we see that the left side is zero at x = −3.36523 as well as two complex values, but we can ignore complex values for interval of validity computations. Finally, a graph of the quantity under the radical is shown below.

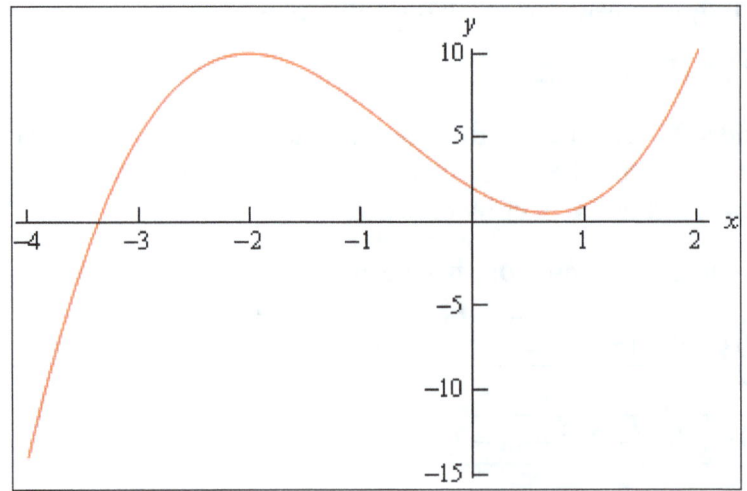

So, in order to get real solutions we will need to require x≥−3.36523 because this is the range of x's for which the quantity is positive. Notice as well that this interval also contains the value of xx that is in the initial condition as it should.

Therefore, the interval of validity of the solution is x≥−3.36523.

Here is graph of the solution.

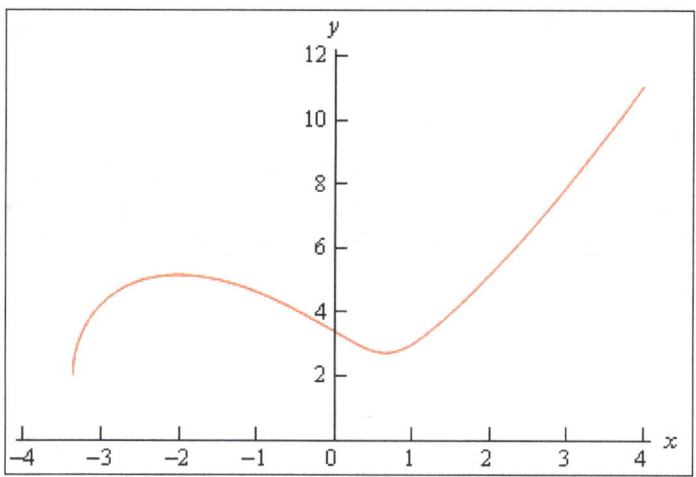

Example: Solve the following IVP and find the interval of validity of the solution.

$$y' = \frac{xy^3}{\sqrt{1+x^2}} \quad y(0) = -1$$

Solution: First separate and then integrate both sides.

$$y^{-3}dy = x(1+x^2)^{-\frac{1}{2}}dx$$

$$\int y^{-3}dy = \int x(1+x^2)^{-\frac{1}{2}}dx$$

$$-\frac{1}{2y^2} = \sqrt{1+x^2} + c$$

Apply the initial condition to get the value of c.

$$-\frac{1}{2} = \sqrt{1} + c \quad c = -\frac{3}{2}$$

The implicit solution is then,

$$-\frac{1}{2y^2} = \sqrt{1+x^2} - \frac{3}{2}$$

Now solve for y(x).

$$\frac{1}{y^2} = 3 - 2\sqrt{1+x^2}$$

$$y^2 = \frac{1}{3 - 2\sqrt{1 + x^2}}$$

$$y(x) = \pm \frac{1}{\sqrt{3 - 2\sqrt{1 + x^2}}}$$

Reapplying the initial condition shows us that the "−" is the correct sign. The explicit solution is then,

$$y(x) = -\frac{1}{\sqrt{3 - 2\sqrt{1 + x^2}}}$$

Let's get the interval of validity. That's easier than it might look for this problem. First, since $1 + x^2$ ≥ 0 the "inner" root will not be a problem. Therefore, all we need to worry about is division by zero and negatives under the "outer" root. We can take care of both by requiring:

$$3 - 2\sqrt{1 + x^2} > 0$$
$$3 > 2\sqrt{1 + x^2}$$
$$9 > 4(1 + x^2)$$
$$\frac{9}{4} > 1 + x^2$$
$$\frac{5}{4} > x^2$$

We were able to square both sides of the inequality because both sides of the inequality are guaranteed to be positive in this case. Finally solving for x we see that the only possible range of x's that will not give division by zero or square roots of negative numbers will be,

$$-\frac{\sqrt{5}}{2} < x < \frac{\sqrt{5}}{2}$$

and nicely enough this also contains the initial condition x=0. This interval is therefore our interval of validity.

Here is a graph of the solution.

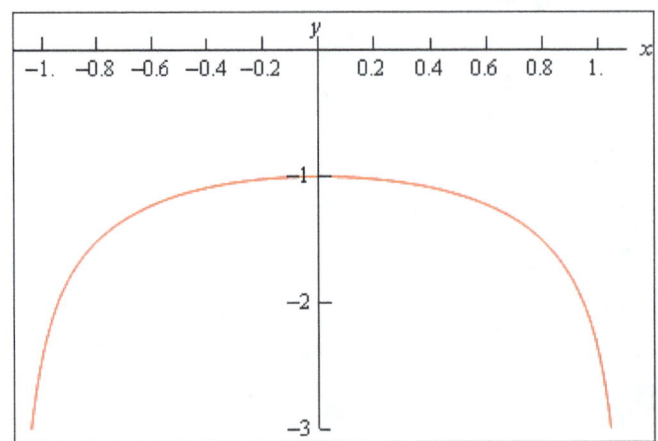

Example: Solve the following IVP and find the interval of validity of the solution.

$$y' = e^{-y}(2x - 4) \qquad y(5) = 0$$

Solution: This differential equation is easy enough to separate, so let's do that and then integrate both sides.

$$
\begin{aligned}
e^y\, dy &= (2x - 4)dx \\
\int e^y\, dy &= \int (2x - 4)dx \\
e^y &= x^2 - 4x + c
\end{aligned}
$$

Applying the initial condition gives:

$$1 = 25 - 20 + c \qquad c = -4$$

This then gives an implicit solution of:

$$e^y = x^2 - 4x - 4$$

We can easily find the explicit solution to this differential equation by simply taking the natural log of both sides.

$$y(x) = \ln(x^2 - 4x - 4)$$

Finding the interval of validity is the last step that we need to take. Recall that we can't plug negative values or zero into a logarithm, so we need to solve the following inequality:

$$x^2 - 4x - 4 > 0$$

The quadratic will be zero at the two points $x = 2 \pm 2\sqrt{2}$ A graph of the quadratic shows that there are in fact two intervals in which we will get positive values of the polynomial and hence can be possible intervals of validity.

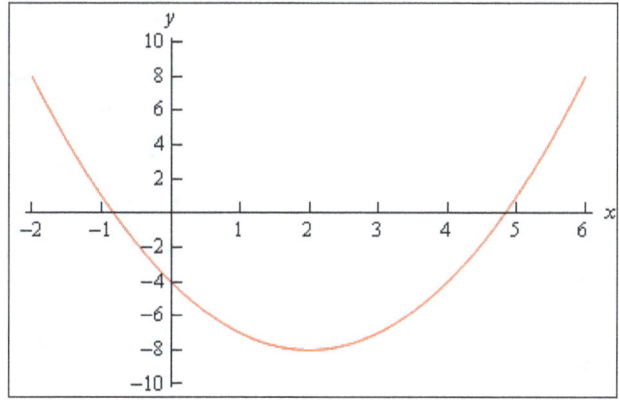

So, possible intervals of validity are:

$$\infty < x < 2 - 2\sqrt{2}$$
$$2 + 2\sqrt{2} < x < \infty$$

From the graph of the quadratic we can see that the second one contains x = 5, the value of the independent variable from the initial condition. Therefore, the interval of validity for this solution is:

$$2 + 2\sqrt{2} < x < \infty$$

Here is a graph of the solution.

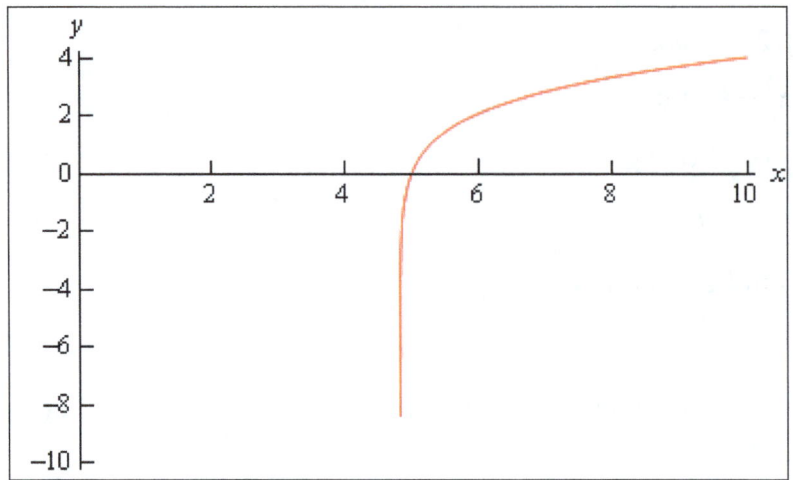

Example: Solve the following IVP and find the interval of validity for the solution.

$$\frac{dr}{d\theta} = \frac{r^2}{\theta} \qquad r(1) = 2$$

Solution: This is actually a fairly simple differential equation to solve. We're doing this one mostly because of the interval of validity.

So, get things separated out and then integrate.

$$\frac{1}{r^2}\,dr = \frac{1}{\theta}\,d\theta$$

$$\int \frac{1}{r^2}\,dr = \int \frac{1}{\theta}\,d\theta$$

$$-\frac{1}{r} = \ln|\theta| + c$$

Now, apply the initial condition to find c.

$$-\frac{1}{2} = \ln(1) + c \qquad c = -\frac{1}{2}$$

So, the implicit solution is then,

$$-\frac{1}{r} = \ln|\theta| - \frac{1}{2}$$

Solving for r gets us our explicit solution:

$$r = \cfrac{1}{\dfrac{1}{2} - \ln|\theta|}$$

Now, there are two problems for our solution here. First, we need to avoid θ=0 because of the natural log. Notice that because of the absolute value on the θ we don't need to worry about θ being negative. We will also need to avoid division by zero. In other words, we need to avoid the following points.

$$\frac{1}{2} - \ln|\theta| = 0$$

$$\ln|\theta| = \frac{1}{2} \quad \text{exponentiate both sides}$$

$$|\theta| = e^{\frac{1}{2}}$$

$$\theta = \pm\sqrt{e}$$

So, these three points break the number line up into four portions, each of which could be an interval of validity.

$$-\infty < \theta - \sqrt{e}$$
$$-\sqrt{e} < \theta < 0$$
$$0 < \theta < \sqrt{e}$$
$$\sqrt{e} < \theta < \infty$$

The interval that will be the actual interval of validity is the one that contains $\theta = 1$. Therefore, the interval of validity is $0 < \theta < \sqrt{e}$.

Here is a graph of the solution.

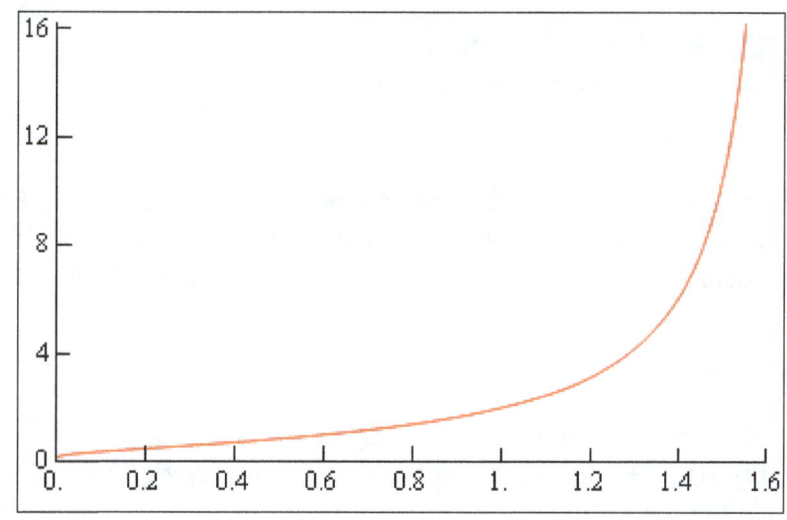

Example: Solve the following IVP.

$$\frac{dy}{dt} = e^{y-t}\sec(y)(1+t^2) \qquad y(0) = 0$$

Solution: This problem will require a little work to get it separated and in a form that we can integrate, so let's do that first.

$$\frac{dy}{dt} = \frac{e^y e^{-t}}{\cos(y)}(1+t^2)$$

$$e^{-y}\cos(y)dy = e^{-t}(1+t^2)dt$$

Now, with a little integration by parts on both sides we can get an implicit solution.

$$\int e^{-y}\cos(y)dy = \int e^{-t}(1+t^2)dt$$

$$\frac{e^{-y}}{2}(\sin(y) - \cos(y)) = -e^{-t}(t^2+2t+3) + c$$

Applying the initial condition gives.

$$\frac{1}{2}(-1) = -(3) + c \qquad c = \frac{5}{2}$$

Therefore, the implicit solution is.

$$\frac{e^{-y}}{2}(\sin(y) - \cos(y)) = -e^{-t}(t^2+2t+3) + \frac{5}{2}$$

It is not possible to find an explicit solution for this problem and so we will have to leave the solution in its implicit form. Finding intervals of validity from implicit solutions can often be very difficult so we will also not bother with that for this problem.

As this last example showed it is not always possible to find explicit solutions.

Homogeneous Equations

Homogeneous Functions

A homogeneous function is one that exhibits multiplicative scaling behavior i.e. if all of its arguments are multiplied by a factor, then the value of the function is multiplied by some power of that factor. Mathematically, we can say that a function in two variables $f(x,y)$ is a homogeneous function of degree n if:

$$f(\alpha x, \alpha y) = \alpha^n f(x,y)$$

where α is a real number. For example, if given $f(x,y,z) = x^2 + y^2 + z^2 + xy + yz + zx$. We can note that $f(\alpha x, \alpha y, \alpha z) = (\alpha x)^2 + (\alpha y)^2 + (\alpha z)^2 + \alpha x.\alpha y + \alpha y.\alpha z + \alpha z.\alpha x = \alpha^2 f(x,y,z)$. Thus, given $f(x,y,z)$ is a homogeneous function of degree 2.

Homogeneous Differential Equation of the First Order

A first-order differential equation, that may be easily expressed as:

$$\frac{dy}{dx} = f(x, y)$$

is said to be a homogeneous differential equation if the function on the right-hand side is homogeneous in nature, of degree = 0. This implies that for any real number α:

$$f(\alpha x, \alpha y) = \alpha^0 f(x, y)$$
$$= f(x, y)$$

An alternate form of representation of the differential equation can be obtained by rewriting the homogeneous function on the right-hand side in terms of two homogeneous functions M(x,y) and N(x,y) of the same degree:

$$\frac{dy}{dx} = -\frac{M(x, y)}{N(x, y)}$$
$$N(x, y)dy = -M(x, y)dx$$
$$M(x, y)dx + N(x, y)dy = 0$$

Another representation is possible, which is a bit non-trivial to prove and is only valid for the case of the first order. It is shown below:

$$\frac{dy}{dx} = f(\frac{x}{y})$$

Now that you've learnt to identify the homogeneous differential equations, let us look at the general method for solving such equations.

Methods of Solving a Homogeneous Differential Equation

- Introduce a new dependent variable $v = \frac{y}{x}$. It would involve substituting y=vx in the differential equation.

- The differential equation now becomes:

$$x\frac{dv}{dx} + v = f(\frac{1}{v})$$

- Solve the above differential equation by the variables separation method.

- Substitute the expression obtained for v back in y=vx to obtain the general solution to the differential equation.

Equations Reducible to the Homogeneous Form

The first-order differential equations of the form:

$$\frac{dy}{dx} = f(\frac{ax + by + c}{kx + ly + m})$$

where a, b, c, k, l, m are constants is reducible to Homogeneous Form. First, let us note how does it even comes close to the homogeneous form. On rewriting the equation:

$$\frac{dy}{dx} = f\left(\frac{a+b\dfrac{y}{x}+\dfrac{c}{x}}{k+l\dfrac{y}{x}+\dfrac{m}{x}}\right)$$

This clearly is the Homogeneous Form itself, if c = m = 0. In order to do just that, we shift the co-ordinates i.e. introduce new variables in the system:

X = x − α, Y = y − β

where α and β are some constants whose values we'll determine from the condition c = m = 0. Following this process, the equation will reduce to homogeneous form which then can be directly solved by the substitution Y=vX. Although, in the end, we must back-substitute Y in the equation Y = y − β to get the final solution.

Example: $(2x^3 + y^3)dx - 3xy^2 dy = 0$

Solution: The given differential equation is a homogeneous differential equation of the first order since it has the form $M(x, y)dx + N(x, y)dy = 0$ where M(x,y) and N(x,y) are homogeneous functions of the same degree = 3 in this case.

Here, $M(x, y) = 2x^3 + y^3$ and $N(x, y) = -3xy^2$ To solve, we first rearrange the differential equation in the following format:

$$\frac{dy}{dx} = \frac{2x^3 + y^3}{3xy^2}$$

To see that the equation is homogeneous, we can also see that the right-side can be converted to a function of the form $f(\frac{y}{x})$ as:

$$\frac{2x^3}{3xy^2} + \frac{y^3}{3xy^2} = \frac{2}{3(\frac{y}{x})^2} + \frac{1}{3}(\frac{y}{x})$$

Anyway, to solve now we proceed with the substitution y=vx where x is the new dependent variable. Then the differential equation in the form gets converted into:

$$v + x\frac{dv}{dx} = \frac{2x^3 + y^3}{3xy^2}$$

$$= \frac{2x^3 + (vx)^3}{3x(vx)^2}$$

$$= \frac{2 + v^3}{3v^2}$$

Then the differential equation can be converted to variables-separable form as:

$$x\frac{dv}{dx} = \frac{2+v^3}{3v^2} - v$$

$$= \frac{2-2v^3}{3v^2}$$

$$\frac{3v^2}{v^3-1}\frac{dv}{dx} = -2x$$

On integrating with respect to x, the general solution is found out to be

$$\int \frac{3v^2}{v^3-1}\frac{dv}{dx}dx = \int \frac{-2}{x}dx$$

$$ln[mod(v^3-1)] = -2lnx + c$$

$$ln[mod(v^3-1)x^2] = c$$

$$[mod(v^3-1)]x^2 = e^c = C$$

This is the implicit form of the general solution for v(x). Now let us find y(x) by back-substituting $v = \frac{y}{x}$.

$$[mod((\frac{y}{x})^3-1)]x^2 = C$$

$$(\frac{y}{x})^3-1)x^2 = \pm C$$

$$y^3-x^3 = \pm Cx$$

which is the final solution. It is actually a one-parameter family of curves which satisfies the homogeneous differential equation.

Example: Solve the equation,

$$(2x-y+1)dx - (x-2y-1)dy = 0$$

Solution: The given differential equation can be rearranged to get,

$$\frac{dy}{dx} = \frac{(2x-y+1)}{(x-2y-1)}$$

This is clearly the form that is reducible to the homogeneous differential equation form. Let us introduce two new variables:

$$X = x - \alpha, Y = y - \beta$$

where α, β are unknown constants. Thus, in our equation, we must proceed with the substitution:

$x = X + \alpha, y = Y + \beta$

$dx = dX, dy = dY$

We can then get the resultant differential equation as:

$$\frac{dY}{dX} = \frac{(2(X+\alpha)-(Y+\beta)+1)}{((X+\alpha)-2(Y+\beta)-1)}$$

$$\frac{dY}{dX} = \frac{(2X-Y+(2\alpha-\beta+1))}{(X-2Y+(\alpha-2\beta-1))}$$

In order for the differential equation to be homogeneous, the terms $(2\alpha-\beta+1)$ and $(\alpha-2\beta-1)$ must be identically equal to zero. Thus we have two simultaneous linear equations in two unknowns (α and β) as:

$2\alpha - \beta + 1 = 0$

$\alpha - 2\beta - 1 = 0$

These can be easily solved to get α = -1, and β = -1. On using these values, we will get the resultant differential equation as:

$$\frac{dY}{dX} = \frac{(2X-Y)}{(X-2Y)}$$

This is clearly Homogeneous in nature and we can proceed by substituting Y = vX

$$v + X\frac{dv}{dX} = \frac{2X-(vX)}{X-2(vX)}$$

$$= \frac{2-v}{1-2v}$$

$$X\frac{dv}{dX} = \frac{2-v}{1-2v} - v$$

$$= \frac{2-2v+2v^2}{1-2v}$$

$$= \frac{2(v^2-v+1)}{1-2v}$$

$$\frac{1-2v}{2(v^2-v+1)}\frac{dv}{dX} = \frac{1}{X}$$

Integrate with respect to X,

$$\int \frac{1-2v}{2(v^2-v+1)}\frac{dv}{dX}dX = \int \frac{1}{X}dX$$

$$ln(v^2-v+1)^{-\frac{1}{2}} = lnX + c$$

$$ln(\frac{(v^2-v+1)^{-\frac{1}{2}}}{X}) = c$$

$$(v^2-v+1)^{\frac{1}{2}}X = \frac{1}{e^c} = C$$

$$(v^2-v+1)X^2 = C'$$

Now let us back-substitute to get Y,

$$\left(\frac{Y}{X}\right)^2 - \frac{Y}{X} + 1)X^2 = C'$$

Finally, we must remember to change the variables X and Y back to the original variables x and y. We can then get the general solution, in the implicit form, as:

$$\left(\left(\frac{(y+1)}{(x+1)}\right)^2 - \frac{(y+1)}{(x+1)} + 1\right)(x+1)^2 = C'$$

This concludes our discussion on this very.

Exact Differential Equations

A differential equation is exact when is a total derivative of a function called potential function. Exact equations are simple to integrate—any potential function must be constant. The solutions of the differential equation define level surfaces of a potential function.

A semi-exact differential equation is an equation that is not exact but it can be transformed into an exact equation after multiplication by a function, called an integrating factor. An integrating factor converts a non-exact equation into an exact equation. Linear equations are a particular case of semi-exact equations. The integrating factor of a linear equation transforms it into a total derivative—hence, an exact equation. We now generalize this idea to a class of nonlinear equations.

Exact Equations

A differential equation is exact if certain parts of the differential equation have matching partial derivatives. We use this definition because it is simple to check in concrete examples.

An exact differential equation for y is:

$$N(t, y)\, y' + M(t, y) = 0$$

where the functions N and M satisfy:

$$\partial_t N(t, y) = \partial_y M(t, y)$$

Remark: The functions N, M depend on t, y, and we use the notation for partial derivatives:

$$\partial_t N = \frac{\partial N}{\partial t}, \qquad \partial_y M = \frac{\partial M}{\partial y}.$$

In the definition above, the letter y has been used both as the unknown function (in the first equation), and as an independent variable (in the second equation).

Example: Show whether a separable equation $h(y)\, y'(t) = g(t)$ is exact or not.

Solution: If we write the equation as $h(y)\, y' - g(t)=0$, then:

$$\left.\begin{array}{l} N(t, y) = h(y) \Rightarrow \partial_t N(t, y) = 0, \\ M(t, y) = g(t) \Rightarrow \partial_y M(t, y) = 0, \end{array}\right\} \Rightarrow \partial_t N(t, y) = \partial_y M(t, y),$$

hence every separable equation is exact.

Example: Show whether the linear differential equation below is exact or not,

$$y'(t) = a(t)\, y(t) + b(t), \qquad a(t) \neq 0.$$

Solution: We first find the functions N and M rewriting the equation as follows,

$$y' + a(t)y - b(t) = 0 \Rightarrow N(t, y) = 1, \; M(t, y) = -a(t)\, y - b(t).$$

Let us check whether the equation is exact or not,

$$\left.\begin{array}{ll} N(t, y) = 1 & \Rightarrow \partial_t N(t, y) = 0, \\ M(t, y) = -a(t)y - b(t) & \Rightarrow {}_y M(t, y) = -a(t), \end{array}\right\} \Rightarrow \partial_t N(t, y) \neq \partial_y M(t, y).$$

So, the differential equation is not exact.

The following examples show that there are exact equations which are not separable.

Example: Show whether the differential equation below is exact or not,

$$2ty\, y' + 2t + y^2 = 0.$$

Solution: We first identify the functions N and M. This is simple in this case, since,

$$(2ty)\, y' + \left(2t + y^2\right) = 0 \Rightarrow N(t, y) = 2ty, \; M(t, y) = 2t + y^2$$

The equation is indeed exact, since,

$$\left.\begin{array}{l} N(t, y) = 2ty \Rightarrow \partial_t N(t, y) = 2y, \\ M(t, y) = 2t + y^2 \Rightarrow \partial_y M(t, y) = 2y, \end{array}\right\} \Rightarrow \partial_t N(t, y) = \partial_y M(t, y).$$

Therefore, the differential equation is exact.

Example: Show whether the differential equation below is exact or not,

$$\sin(t) y' + t^2 e^y\, y' - y' = -y \cos(t) - 2te^y ..$$

Solution: We first identify the functions N and M by rewriting the equation as follows,

$$\left(\sin(t) + t^2 e^y - 1\right) y' + \left(y\cos(t) + 2te^y\right) = 0$$

we can see that:

$$N(t, y) = \sin(t) + t^2 e^y - 1 \Rightarrow \partial_t N(t, y) = \cos(t) + 2te^y,$$
$$M(t, y) = y \cos(t) + 2te^y \Rightarrow \partial_y M(t, y) = \cos(t) + 2te^y,$$

Therefore, $\partial_t N(t, y) = \partial_y M(t, y)$, and the equation is exact.

Solving Exact Equations

Exact differential equations can be rewritten as a total derivative of a function, called a potential function. Once they are written in such way they are simple to solve.

Theorem of Exact Equations: If the differential equation:

$$N(t, y) y' + M(t, y) = 0$$

is exact, then it can be written as:

$$\frac{d\psi}{dt}(t, y(t)) = 0,$$

where ψ is called a potential function and satisfies:

$$N = \partial_y \psi, \quad M = \partial_t \psi.$$

Therefore, the solutions of the exact equation are given in implicit form as:

$$\psi(t, y(t)) = c, \quad c \in \mathbb{R}$$

The condition $\partial_t N(t, y) = \partial_y M$ M is equivalent to the existence of a potential function— result proven by Henri Poincaré around 1880.

Theorem of Poincaré: Continuously differentiable functions N, M, on t, y, satisfy,

$$\partial_t N(t, y) = \partial_y M(t, y)$$

if there is a twice continuously differentiable function ψ, depending on t, y such that:

$$\partial_y \psi(t, y) = N(t, y), \quad \partial_t \psi(t, y) = M(t, y).$$

Remarks:

A differential equation defines the functions N and M. The exact condition in equation $\partial_t N(t, y) = \partial_y M(t, y)$ is equivalent to the existence of ψ, related to N and M through Equation $\partial_y \psi(t, y) = N(t, y)$, $\partial_t \psi(t, y) = M(t, y)$.

If we recall the definition of the gradient of a function of two variables, $\nabla \psi = \langle \partial_t \psi, \partial_y \psi \rangle$, then the equations in $\partial_y \psi(t, y) = N(t, y)$, $\partial_t \psi(t, y) = M(t, y)$ say that $\nabla \psi = \langle M, N \rangle$.

Proof of Lemma

($\Rightarrow$) It is not given.

($\Leftarrow$) We assume that the potential function ψ is given and satisfies:

$$N = \partial_y \psi, \quad M = \partial_t \psi.$$

Recalling that ψ is twice continuously differentiable, hence $\partial_t \partial_y \psi = \partial_y \partial_t \psi$, then we have:

$$\partial_t N = \partial_t \partial_y \psi = \partial_y \partial_t \psi = \partial_y M.$$

In our next example we verify that a given function ψ is a potential function for an exact differential equation. We also show that the differential equation can be rewritten as a total derivative of this potential function.

Example: (Verification of a Potential) - Show that the differential equation,

$$2ty \, y' + 2t + y^2 = 0.$$

is the total derivative of the potential function $\psi(t, y) = t^2 + ty^2$.

Solution: We use the chain rule to compute the t derivative of the potential function ψ evaluated at the unknown function y,

$$\frac{d}{dt}\psi(t, y(t)) = (\partial_y \psi)\frac{dy}{dt} + (\partial_t \psi)$$
$$= (2ty)y' + (2t + y^2).$$

So the differential equation is the total derivative of the potential function. To get this result we used the partial derivatives:

$$\partial_y \psi = 2ty = N, \quad \partial_t \psi = 2t + y^2 = M.$$

Exact equations always have a potential function ψ, and this function is not difficult to compute—we only need to integrate Equation $\partial_y \psi(t, y) = N(t, y)$, $\partial_t \psi(t, y) = M(t, y)$. Having a potential function of an exact equation is essentially the same as solving the differential equation, since the integral curves of ψ define implicit solutions of the differential equation.

Proof of Theorem: The differential equation in $N(t, y)\, y' + M(t, y) = 0$ is exact, then Poincaré Theorem implies that there is a potential function ψ such that:

$$N = \partial_y \psi, \quad M = \partial_t \psi.$$

Therefore, the differential equation is given by:

$$
\begin{aligned}
0 &= N(t, y)\, y'(t) + M(t, y) \\
&= (\partial_y \psi(t, y))y' + (\partial_t \psi(t, y)) \\
&= \frac{d}{dt}\psi(t, y(t)),
\end{aligned}
$$

where in the last step we used the chain rule. Recall that the chain rule says:

$$\frac{d}{dt}\psi(t, y(t)) = (\partial_y \psi)\frac{dy}{dt} + (\partial_t \psi).$$

So, the differential equation has been rewritten as a total t-derivative of the potential function, which is simple to integrate,

$$\frac{d}{dt}\psi(t, y(t)) = 0 \Rightarrow \psi(t, y(y)) = c,$$

where c is an arbitrary constant. This establishes the Theorem.

Example: (Calculation of a Potential) - Find all solutions y to the differential equation,

$$2ty\, y' + 2t + y^2 = 0.$$

Solution: The first step is to verify whether the differential equation is exact. We know the answer, the equation is exact,

$$
\left.
\begin{aligned}
N(t, y) &= 2ty & &\Rightarrow \partial_t N(t, y) = 2y, \\
M(t, y) &= 2t + y^2 & &\Rightarrow \partial_y M(t, y) = 2y.
\end{aligned}
\right\} \Rightarrow \partial_t N(t, y) = \partial_y M(t, y).
$$

Since the equation is exact, Lemma implies that there exists a potential function ψ satisfying the equations:

$$
\begin{aligned}
\partial_y \psi(t, y) &= N(t, y), \\
\partial_t \psi(t, y) &= M(t, y).
\end{aligned}
$$

Let us compute ψ. Integrate equation $\partial_y \psi(t, y) = N(t, y)$, in the variable y keeping the variable t constant,

$$\partial_y \psi(t, y) = 2ty \Rightarrow \psi(t, y) = \int 2ty\, dy + g(t),$$

where g is a constant of integration on the variable y, so g can only depend on t. We obtain:

$$\psi(t, y) = ty^2 + g(t).$$

Introduce into equation $\partial_t \psi(t,y) = M(t,y)$ the expression for the function ψ in Equation above, that is,

$$y^2 + g'(t) = \partial_t \psi(t, y) = M(t, y) = 2t + y^2 \Rightarrow g'(t) = 2t$$

Integrate in t the last equation above, and choose the integration constant to be zero,

$$g(t) = t^2.$$

We have found that a potential function is given by:

$$\psi(t,y) = ty^2 + t^2.$$

Therefore, theorem of exact equation implies that all solutions y satisfy the implicit equation:

$$ty^2(t) + t^2 = c,$$

For any $c \in R$. The choice $g(t) = t^2 + c_0$ only modifies the constant c.

Remark: An exact equation and its solutions can be pictured on the graph of a potential function. This is called a geometrical interpretation of the exact equation. We saw that an exact equation $N y' + M = 0$ can be rewritten as $d\psi / dt = 0$. The solutions of the differential equation are functions y such that $\psi(t, y(t)) = c$, hence the solutions define level curves of the potential function. Given a level curve, the vector $r(t) = \langle t, y(t) \rangle$, which belongs to the ty-plane, points to the level curve, while its derivative $r'(t) = \langle 1, y'(t) \rangle$ is tangent to the level curve. Since the gradient vector $\nabla \psi = \langle M, N \rangle$ is perpendicular to the level curve,

$$r' \perp \nabla \psi \Leftrightarrow r' \cdot \nabla \psi = 0 \Leftrightarrow M + N y' = 0.$$

We wanted to remark that the differential equation can be thought as the condition $r' \perp \nabla \psi$.

As an example, consider the differential equation:

$$2y\, y' + 2t = 0,$$

which is separable, so it is exact. A potential function is:

$$\psi = t^2 + y^2,$$

a paraboloid shown in Figure. Solutions y are defined by the equation $t^2 + y^2 = c$, which are level curves of ψ for c > 0. The graph of a solution is shown on the ty-plane, given by:

$$y(t) = \pm\sqrt{c - t^2}.$$

As we said above, the vector $r(t) = \langle t, y(t) \rangle$ points to the solution's graph while its derivative $r'(t) = \langle 1, y'(t) \rangle$ is tangent to the level cuve. We also know that the gradient vector $\nabla \psi = \langle 2t, 2y \rangle$ is perpendicular to the level curve. The condition:

$$r' \perp \nabla \psi \Rightarrow r' \cdot \nabla \psi = 0,$$

is precisely the differential equation,

$$2t + 2y\,y' = 0.$$

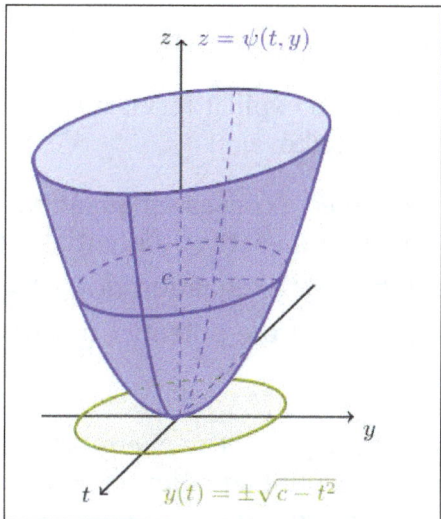

Potential ψ with level curve $\psi = c$ defines a solution y on the ty-plane.

Example: (Calculation of a Potential). Find all solutions y to the equation,

$$\sin(t)y' + t^2 e^y\, y' - y' + y\cos(t) + 2te^y = -3t^2 = 0.$$

Solution: The first step is to verify whether the differential equation is exact.

$$N(t, y) = \sin(t) + t^2 e^y - 1 \quad \Rightarrow \quad \partial_t N(t, y) = \cos(t) + 2te^y,$$
$$M(t, y) = y\cos(t) + 2te^y - 3t^2 \quad \Rightarrow \quad \partial_y M(t, y) = \cos(t) + 2te^y.$$

So, the equation is exact. Poincaré Theorem says there is a potential function ψ satisfying

$$\partial_y \psi(t, y) = N(t, y),\ \partial_t \psi(t, y) = M(t, y).$$

To compute ψ we integrate on y the equation $\partial_y \psi = N$ keeping t constant,

$$\partial_y \psi(t, y) = \sin(t) + t^2 e^y - 1 \Rightarrow \psi(t, y) = \int (\sin(t) + t^2 e^y - 1)\,dy + g(t)$$

where g is a constant of integration on the variable y, so g can only depend on t. We obtain,

$$\psi(t, y) = y\sin(t) + t^2 e^y - y + g(t).$$

Now introduce the expression above for ψ in the second equation in Equation $\partial_y \psi(t, y) = N(t, y),\ \partial_t \psi(t, y) = M(t, y),$

$$y\cos(t) + 2te^y + g'(t) = \partial_t \psi(t, y) = M(t, y) = y\cos(t) + 2te^y - 3t^2 \Rightarrow g'(t) = -3t^2.$$

The solution is $g(t) = -t^3 + c_0$, with c_0 a constant. We choose $c_0 = 0$, so $g(t) = -t^3$. We found g, so we have the complete potential function,

$$\psi(t, y) = y\, sin(t) + t^2 e^y - y - t^3.$$

Theorem of Exact Equation implies that any solution y satisfies the implicit equation:

$$y(t)\, sin(t) + t^2 e^y(t) - y(t) - t^3 = c.$$

The solution y above cannot be written in explicit form. If we choose the constant $c_0 \neq 0$ in g(t) $= -t^3 + c_0$, we only modify the constant c above.

Remark: A potential function is also called a conserved quantity. This is a reasonable name, since a potential function evaluated at any solution of the differential equation is constant along the evolution. This is yet another interpretation of the equation $d\psi/dt = 0$, or its integral $\psi(t, y(t)) = c$. If we call $c = \psi_0 = \psi(0, y(0))$, the value of the potential function at the initial conditions, then $\psi(t, y(t)) = \psi_0$.

Conserved quantities are important in physics. The energy of a moving particle is a famous conserved quantity. In that case the differential equation is Newton's second law of motion, mass times acceleration equals force. One can prove that the energy E of a particle with position function y moving under a conservative force is kept constant in time. This statement can be expressed by $E(t, y(t), y'(t)) = E_0$, where E_0 is the particle's energy at the initial time.

Semi-Exact Equations: Sometimes a non-exact differential equation can be rewritten as an exact equation. One way this could happen is multiplying the differential equation by an appropriate function. If the new equation is exact, the multiplicative function is called an integrating factor.

A semi-exact differential equation is a non-exact equation that can be transformed into an exact equation after a multiplication by an integrating factor.

Example: Linear differential equations $y' = a(t)y + b(t)$ are semi-exact.

Solution: We first show that linear equations $y' = ay + b$ with $a \neq 0$ are not exact. If we write them as:

$$y' - a\,y - b = 0 \;\Rightarrow\; Ny' + M = 0 \text{ with } N = 1,\; M = -a\,y - b.$$

Therefore,

$$\partial_t N = 0,\; \partial_y M = -a \;\Rightarrow\; \partial_t N \neq \partial_y M.$$

We now show that linear equations are semi-exact. Let us multiply the linear equation by a function μ, which depends only on t,

$$\mu(t)\, y' - a(t)\, \mu(t)\, y - \mu(t)\, b(t) = 0,$$

where we emphasized that μ, a, b depend only on t. Let us look for a particular function μ that makes the equation above exact. If we write this equation as $\tilde{N} y' + \tilde{M} = 0$, then:

$$\tilde{N}(t, y) = \mu, \quad \tilde{M}(t, y) = -a\mu y - \mu b.$$

We now check the condition for exactness,

$$\partial_t \tilde{N} = \mu', \quad \partial_y \tilde{M} = - a \mu,$$

and we get that:

$$\left.\begin{array}{c} \partial_t \tilde{N} = \partial_y \tilde{M} \\ \text{the equation is exact} \end{array}\right\} \Leftrightarrow \left\{\begin{array}{l} \mu' = -a\mu \\ \mu \text{ is an integrating factor.} \end{array}\right.$$

Therefore, the linear equation $y' = ay + b$ is semi-exact, and the function that transforms it into an exact equation is $\mu(t) = e^{-A(t)}$, where $A(t) = \int a(t)dt$,

Now we generalize this idea to nonlinear differential equations.

Theorem: If the equation,

$$N(t, y) \, y' + M(t, y) = 0$$

is not exact, with $\partial_t N \neq \partial_y M$, the function $N \neq 0$, and where the function h defined as"

$$h = \frac{\partial_y M - \partial_t N}{N}$$

depends only on t, not on y, then the equation below is exact,

$$(e^H N) \, y' + (e^H M) = 0,$$

where H is an antiderivative of h,

$$H(t) = \int h(t) \, dt.$$

Remarks:

- The function $\mu(t) = e^{H(t)}$ is called an integrating factor.

- Any integrating factor μ is solution of the differential equation

  $\mu'(t) = h(t) \, \mu(t).$

- Multiplication by an integrating factor transforms a non-exact equation

  $N y' + M = 0$

  into an exact equation.

  $(\mu N) \, y' + (\mu M) = 0.$

This is exactly what happened with linear equations.

Verification Proof of Previous Theorem: We need to verify that the equation is exact,

$$(e^H N) \, y' + (e^H M) = 0 \implies \tilde{N}(t, y) = e^{H(t)} N(t, y), \quad \tilde{M}(t, y) = e^{H(t)} M(t, y).$$

We now check for exactness, and let us recall $\partial t(e^H) = (e^H)' = h\, e^H$, then:

$$\partial_t \tilde{N} = h\, e^H\, N + e^H\, \partial_t N, \quad \partial_y \tilde{M} = e^H\, \partial_y M .$$

Let us use the definition of h in the first equation above,

$$\partial_t \tilde{N} = e^H \left(\frac{(\partial_y M - \partial_t N)}{N} N + \partial_t N \right) = e^H \partial_y M = \partial_y \tilde{M}.$$

So the equation is exact. This establishes the Theorem.

Constructive Proof of Theorem: The original differential equation:

$$N y' + M = 0$$

is not exact because $\partial_t N \neq \partial_y M$. Now multiply the differential equation by a nonzero function μ that depends only on t,

$$(\mu N)\, y' + (\mu M) = 0.$$

We look for a function μ such that this new equation is exact. This means that μ must satisfy the equation:

$$\partial_t (\mu N) = \partial_y (\mu M).$$

Recalling that μ depends only on t and denoting $\partial_t \mu = \mu'$, we get:

$$\mu' N + \mu\, \partial_t N = \mu\, \partial_y M \;\Rightarrow\; \mu' N = \mu\, (\partial_y M - \partial_t N).$$

So the differential equation in $(\mu N)\, y' + (\mu M) = 0$ is exact if holds:

$$\mu' = \left(\frac{\partial_y M - \partial_t N}{N} \right) \mu.$$

The solution μ will depend only on t if the function:

$$h(t) = \frac{\partial_y M(t,y) - \partial_t N(t,y)}{N(t,y)}$$

depends only on t. If this happens, as assumed in the hypotheses of the theorem, then we can solve for μ as follows,

$$\mu'(t) = h(t)\, \mu(t) \;\Rightarrow\; \mu(t) = e^{H(t)}, \quad H(t) = \int h(t)\, dt.$$

Therefore, the equation below is exact,

$$(e^H\, N)\, y' + (e^H\, M) = 0.$$

This establishes the Theorem.

Example: Find all solutions y to the differential equation:

$$(t^2 + t\,y)y' + (3t\,y + y^2) = 0.$$

Solution: We first verify whether this equation is exact:

$$N(t, y) = t^2 + ty \qquad \Rightarrow \qquad \partial_t N(t,y) = 2t + y,$$
$$M(t,y) = 3ty + y^2 \qquad \Rightarrow \qquad \partial_y M(t,y) = 3t + 2y,$$

therefore, the differential equation is not exact. We now verify whether the extra condition in the-orem holds, that is, whether the function in equation $h = \dfrac{\partial_y M - \partial_t N}{N}$ is y independent:

$$h = \frac{\partial_y M(t,y) - \partial_t N(t,y)}{N(t,y)}$$
$$= \frac{(3t + 2y) - (2t + y)}{(t^2 + ty)}$$
$$= \frac{(t + y)}{t(t + y)}$$
$$= \frac{1}{t} \;\Rightarrow\; h(t) = \frac{1}{t}.$$

So, the function $h = (\partial_y M - \partial_t N)/N$ is y independent. Therefore, Theorem implies that the non-exact differential equation can be transformed into an exact equation. We need to multiply the differential equation by a function μ solution of the equation:

$$\mu'(t) = h(t)\,\mu(t) \;\Rightarrow\; \frac{\mu'}{\mu} = \frac{1}{t} \;\Rightarrow\; \operatorname{In}(\mu(t)) = \operatorname{In}(t) \;\Rightarrow\; \mu(t) = t,$$

where we have chosen in second equation the integration constant to be zero. Then, multiplying the original differential equation in $(t^2 + t\,y)y' + (3t\,y + y^2) = 0$ by the integrating factor μ we obtain:

$$(3t^2 y + t\,y^2) + (t^3 + t^2 y)y' = 0 = 0.$$

This latter equation is exact, since:

$$\tilde{N}(t, y) = t^3 + t^2 y \qquad \Rightarrow \qquad \partial_t \tilde{N}(t,y) = 3t^2 + 2ty,$$
$$\tilde{M}(t,y) = 3t^2 y + ty^2 \qquad \Rightarrow \qquad \partial_y \tilde{M}(t,y) = 3t^2 + 2ty,$$

so we get the exactness condition $\partial_t \tilde{N} = \partial_y \tilde{M}$. That is, we find the potential function ψ by integrating the equations:

$$\partial_y \psi(t, y) = \tilde{N}(t, y),$$
$$\partial_t \psi(t,y) = \tilde{M}(t,y).$$

From the first equation above we obtain:

$$\partial_y \psi = t^3 + t^2 y \Rightarrow \psi(t, y) = \int (t^3 + t^2 y) dy + g(t).$$

Integrating on the right hand side above we arrive to:

$$\psi(t, y) = t^3 y + \frac{1}{2} t^2 y^2 + g(t).$$

Introduce the expression above for ψ in Equation $\partial_t \psi(t,y) = \tilde{M}(t,y)$,

$$3t^2 y + ty^2 + g'(t) = \partial t \psi(t, y) = \tilde{M}(t, y) = 3t^2 y + ty^2,$$
$$g'(t) = 0.$$

A solution to this last equation is g(t) = 0. So we get a potential function:

$$\psi(t,y) = t^3 + \frac{1}{2} t^2 y^2.$$

All solutions y to the differential equation in equation $(t^2 + t y)y' + (3t y + y^2) = 0$ satisfy the equation:

$$t^3 y(t) + \frac{1}{2} t^2 (y(t))^2 = c_0,$$

Where $c_0 \in \mathbb{R}$ is arbitrary.

We have seen in Example that linear differential equations with a ≠ 0 are not exact.

Example: Use Theorem to find all solutions to the linear differential equation,

$$y' = a(t)y + b(t), \quad a(t) \neq 0.$$

Solution: We first write the linear equation in a way we can identify functions N and M,

$$y' - (a(t) y + b(t)) = 0.$$

We now verify whether the linear equation is exact or not. Actually, we have seen in Example that this equation is not exact, since:

$$N(t,y) = 1 \qquad \Rightarrow \qquad \partial_t N(t,y) = 0,$$
$$M(t,y) = -a(t)y - b(t) \Rightarrow \qquad \partial_y M(t,y) = -a(t).$$

But now we can go further, we can check whether the condition in Theorem holds or not. We compute the function:

$$\frac{\partial_y M(t,y) - \partial_t N(t,y)}{N(t,y)} = \frac{-a(t) - 0}{1} = -a(t)$$

and we see that it is independent of the variable y. Theorem says that we can transform the linear equation into an exact equation. We only need to multiply the linear equation by a function μ, solution of the equation:

$$\mu'(t) = -a(t)\mu(t) \quad \Rightarrow \quad \mu(t) = e^{-A(t)}, \; A(t) = \int a(t)dt.$$

Therefore, the equation below is exact,

$$e^{-A(t)}y' - (a(t)e^{-A(t)}y - b(t)e^{-A(t)}) = 0.$$

This new version of the linear equation is exact, since:

$$\tilde{N}(t,y) = e^{-A(t)} \qquad \Rightarrow \qquad \partial_t \tilde{N}(t,y) = -a(t)e^{-A(t)},$$
$$\tilde{M}(t,y) = -a(t)e^{-A(t)}y - b(t)e^{-A(t)} \quad \Rightarrow \quad \partial_y \tilde{M}(t,y) = -a(t)e^{-A(t)}.$$

Since the linear equation is now exact, the solutions y can be found as we did in the previous examples in this topic. We find the potential function ψ integrating the equations:

$$\partial_y \psi(t,y) = \tilde{N}(t,y),$$
$$\partial_t \psi(t,y) = \tilde{M}(t,y).$$

From the first equation above we obtain:

$$\partial_y \psi = e^{-A(t)} \Rightarrow \psi(t,y) = \int e^{-A(t)}dy + g(t).$$

The integral is simple, since $e^{-A(t)}$ is y independent. We then get:

$$\psi(t,y) = e^{-A(t)}y + g(t).$$

We introduce the expression above for ψ in Equation $\partial_t \psi(t,y) = \tilde{M}(t,y)$,

$$-a(t)e^{-A(t)}y + g'(t) = \partial_t \psi(t,y) = \tilde{M}(t,y) = -a(t)e^{-A(t)}y - b(t)e^{-A(t)},$$
$$g'(t) = -b(t)e^{-A(t)}.$$

A solution for function g is then given by:

$$g(t) = -\int b(t) e^{-A(t)}dt.$$

Having that function g, we get a potential function:

$$\psi(t,y) = e^{-A(t)}y - \int b(t) e^{-A(t)}dt.$$

All solutions y to the linear differential equation in $y' = a(t)y + b(t), \; a(t) \neq 0.$ satisfy the equation:

$$e^{-A(t)}y(t) - \int b(t) e^{-A(t)}dt = c_0,$$

Where $c_0 \in \mathbb{R}$ is arbitrary. This is the implicit form of the solution, but in this case it is simple to find the explicit form too,

$$y(t) = e^{A(t)}\left(c_0 + \int b(t)\, e^{-A(t)} dt\right).$$

This expression agrees with the one in Theorem (Variable coefficients IVP) when we studied linear equations.

The Equation for the Inverse Function

Sometimes the equation for a function y is neither exact nor semi-exact, but the equation for the inverse function y −1 might be. We now try to find out when this can happen. To carry out this study it is more convenient to change a little bit the notation we have been using so far:

- We change the independent variable name from t to x. Therefore, we write differential equations as:

$$N(x, y)\, y' + M(x, y) = 0, \quad y = y(x), \quad y' = \frac{dy}{dx}.$$

- We denote by $x(y)$ the inverse of $y(x)$, that is:

$$x(y_1) = x_1 \iff y(x_1) = y_1.$$

- Recall the identity relating derivatives of a function and its inverse function:

$$x'(t) = \frac{1}{y'(x)}.$$

Our first result says that for exact equations it makes no difference to solve for y or its inverse x. If one equation is exact, so is the other equation.

Theorem: $N y' + M = 0$ is exact $\iff$ $M x' + N = 0$ is exact.

Remark: We will see that for semi-exact equations there is a difference.

Proof of Theorem: Write the differential equation of a function y with values $y(x)$,

$$N(x, y)\, y' + M(x, y) = 0 \text{ and } \partial_x N = \partial_y M.$$

If a solution y is invertible we denote $y^{-1}(y) = x(y)$, and we have the well-known relation

$$x'(y) = \frac{1}{y'(x(y))}.$$

Divide the differential equation above by y' and use the relation above, then we get

$$N(x, y) + M(x, y)\, x' = 0,$$

where now y is the independent variable and the unknown function is x, with values $x(y)$, and the prime means $x' = dx/dy$. The condition for this last equation to be exact is:

$$\partial_y M = \partial_x N,$$

which is exactly the same condition for the equation $N\,y'+M=0$ to be exact. This establishes the Theorem.

Remark: Sometimes, in the literature, the equations $N\,y'+M=0$ and $N+M\,x'=0$ are written together as follows,

$$N\,dy + M\,dx = 0.$$

This equation deserves two comments:

- We do not use this notation here. That equation makes sense in the framework of differential forms, which is beyond the subject of these notes.

- Some people justify the use of that equation outside the framework of differential forms by thinking $y' = \dfrac{dy}{dx}$ as real fraction and multiplying $N\,y'+M=0$ by the denominator,

$$N\frac{dy}{dx} + M = 0 \implies N\,dy + M\,dx = 0.$$

Unfortunately, y' is not a fraction $\dfrac{dy}{dx}$, so the calculation just mentioned has no meaning.

So, if the equation for y is exact, so is the equation for its inverse x. The same is not true for semi-exact equations. If the equation for y is semi-exact, then the equation for its inverse x might or might not be semi-exact. The next result states a condition on the equation for the inverse function x to be semi-exact. This condition is not equal to the condition on the equation for the function y to be semi-exact.

Theorem: If the equation,

$$M\,x'+N = 0$$

is not exact, with $\partial_y M \neq \partial_x N$, the function $M \neq 0$, and where the function ℓ defined as:

$$\ell = -\frac{(\partial_y M - \partial_x N)}{M}$$

depends only on y, not on x, then the equation below is exact,

$$(e^L M)x'+(e^L N) = 0$$

where L is an antiderivative of ℓ,

$$L(y) = \int \ell(y)\,dy.$$

Remarks:

- The function $\mu(y) = e^{L(y)}$ is called an integrating factor.

- Any integrating factor μ is solution of the differential equation:

$$\mu'(y) = \ell(y)\mu(y).$$

- Multiplication by an integrating factor transforms a non-exact equation:

$$M x' + N = 0$$

into an exact equation.

$$(\mu M)x' + (\mu N) = 0.$$

Verification Proof of Theorem: We need to verify that the equation is exact,

$$(e^L M) x' + (e^L N) = 0 \quad \Rightarrow \quad \tilde{M}(x, y) = e^{L(y)} M(x, y), \quad \tilde{N}(x, y) = e^{L(y)} N(x,y).$$

We now check for exactness, and let us recall $\partial_y (e^L) = (e^L)' = \ell e^L$, then:

$$\partial_y \tilde{M} = \ell e^L M + e^L \partial_y M, \quad \partial_x \tilde{N} = e^H \partial_x N.$$

Let us use the definition of ℓ in the first equation above,

$$\partial_y \tilde{M} = e^L \left(-\frac{\partial_y M - \partial_x N}{M} M + \partial_y M \right) = e^L \partial_x N = \partial_x \tilde{N}.$$

So the equation is exact. This establishes the Theorem.

Constructive Proof of Theorem: The original differential equation:

$$M x' + N = 0$$

is not exact because $\partial_y M \neq \partial_x N$. Now multiply the differential equation by a nonzero function μ that depends only on y,

$$(\mu M) x' + (\mu N) = 0.$$

We look for a function μ such that this new equation is exact. This means that μ must satisfy the equation:

$$\partial_y (\mu M) = \partial_x (\mu N).$$

Recalling that μ depends only on y and denoting $\partial_y \mu = \mu'$, we get:

$$\mu' M + \mu \partial_y M = \mu \partial_x N \Rightarrow \mu' M = -\mu (\partial_y M - \partial_x N).$$

So the differential equation $(\mu M) x' + (\mu N) = 0$ is exact if holds:

$$\mu' = -\left(\frac{\partial_y M - \partial_x N}{M} \right) \mu.$$

The solution μ will depend only on y if the function:

$$\ell(y) = -\frac{\partial_y M(x,y) - \partial_x N(x,y)}{M(x,y)}$$

depends only on y. If this happens, as assumed in the hypotheses of the theorem, then we can solve for μ as follows,

$$\mu'(y) = \ell(y)\,\mu(y) \;\Rightarrow\; \mu(y) = e^{L(y)},\; L(y) = \int \ell(y)\,dy.$$

Therefore, the equation below is exact,

$$(e^L\,M)\,x' + (e^L\,N) = 0.$$

This establishes the Theorem.

Example: Find all solutions to the differential equation,

$$(5x\,e^{-y} + 2\,cos(3x))y' + (5\,e^{-y} - 3\,sin(3x)) = 0.$$

Solution: We first check if the equation is exact for the unknown function y, which depends on the variable x. If we write the equation $N\,y' + M = 0$, with $y' = dy/dx$, then:

$$N(x,y) = 5x\,e^{-y} + 2\cos(3x) \;\Rightarrow\; \partial_x N(x,y) = 5\,e^{-y} - 6\sin(3x),$$
$$M(x,y) = 5\,e^{-y} - 3\,\sin(3x) \;\Rightarrow\; \partial_y M(x,y) = -5\,e^{-y}$$

Since $\partial_x N \neq \partial_y M$, the equation is not exact. Let us check if there exists an integrating factor μ that depends only on x. Following Theorem we study the function:

$$h = \frac{(\partial_y M - \partial_x N)}{N} = \frac{-10e^{-y} + 6\sin(3x)}{5xe^{-y} + 2\cos(3x)},$$

which is a function of both x and y and cannot be simplified into a function of x alone. Hence an integrating factor cannot be function of only x.

Let us now consider the equation for the inverse function x, which depends on the variable y. The equation is $M\,x' + N = 0$, with $x' = dx/dy$, where M and N are the same as before,

$$M(x,y) = 5\,e^{-y} - 3\sin(3x) \quad N(x,y) = 5xe^{-y} + 2\cos(3x).$$

We know from Theorem that this equation is not exact. Both the equation for y and equation for its inverse x must satisfy the same condition to be exact. The condition is $\partial_x N = \partial_y M$, but we have seen that this is not true for the equation in this example. The last thing we can do is to check if the equation for the inverse function x has an integrating factor μ that depends only on y. Following Theorem we study the function:

$$\ell = -\frac{(\partial_y M - \partial_x N)}{M} = -\frac{(-10e^{-y} + 6\sin(3x))}{(5e^{-y} - 3\sin(3x))} = 2 \;\Rightarrow\; \ell(y) = 2.$$

The function above does not depend on x, so we can solve the differential equation for μ(y),

$$\mu'(y) = \ell(y)\,\mu(y) \;\Rightarrow\; \mu'(y) = 2\mu(y) \;\Rightarrow\; \mu(y) = \mu_0 e^{2y}.$$

Since μ is an integrating factor, we can choose $\mu_o = 1$, hence $\mu(y) = e^{2y}$. If we multiply the equation for x by this integrating factor we get:

$$e^{2y}(5\,e^{-y} - 3\sin(3x))x' + e^{2y}(5x\,e^{-y} + 2\cos(3x))=0,$$

$$(5\,e^{y} - 3\sin(3x)\,e^{2y})x' +(5x\,e^{y} + 2\cos(3x)\,e^{2y})=0.$$

This equation is exact, because if we write $\tilde{M}x'+ \tilde{N} = 0$, then:

$$\tilde{M}(x,y) = 5e^{y} -3\sin(3x))e^{2y} \quad\Rightarrow\quad \partial_y\tilde{M}(x,y) = 5e^{y} - 6\sin(3x)e^{2y},$$

$$\tilde{N}(x,y) = 5xe^{y} +2\cos(3x)e^{2y} \quad\Rightarrow\quad \partial_x\tilde{N}(x,y) = 5e^{y} - 6\sin(3x)e^{2y},$$

That is $\partial_y\tilde{M} = \partial_x\tilde{N}$. Since the equation is exact, we find a potential function ψ from:

$$\partial_x\psi = \tilde{M}, \quad \partial_y\psi = \tilde{N}.$$

Integrating on the variable x the equation $\partial_x\psi = \tilde{M}$ we get:

$$\psi(x,y) = 5xe^{y} +\cos(3x)e^{2y} +g(y).$$

Introducing this expression for ψ into the equation $\partial_y\psi = \tilde{N}$ we get:

$$5xe^{y} +2\cos(3x)e^{2y} +g'(y) = \partial_y\psi = \tilde{N} = 5xe^{y} +2\cos(3x)e^{2y},$$

Hence $g'(y) = 0$, so we choose g = 0. A potential function for the equation for x is:

$$\psi(x,y) = 5xe^{y} +\cos(3x)e^{2y}.$$

The solutions x of the differential equation are given by:

$$5\,x(y)\,e^{y} + \cos(3x(y))\,e^{2y} = c.$$

Once we have the solution for the inverse function x we can find the solution for the original unknown y, which are given by:

$$5x\,e^{y(x)} + \cos(3x)\,e^{2y(x)} = c.$$

First Order Linear Differential Equations

A linear first order equation is one that can be reduced to a general form:

$$\frac{dy}{dx}+P(x)y = Q(x)$$

where P(x) and Q(x) are continuous functions in the domain of validity of the differential equation. If P(x) or Q(x) is equal to 0, the differential equation can be reduced to a variables separable form which can be easily solved. You can check this for yourselves.

Integrating Factor

To find the solution of the linear first order differential equation as defined above, we must introduce the concept of an integrating factor. An integrating factor is a term, which when multiplied by an expression, converts it to an exact differential i.e. a function which is the derivative of another function. For example: given an expression,

$$y - x\frac{dy}{dx}$$

If we multiply this expression by $\frac{1}{x^2}$ note that it becomes an exact differential,

$$\frac{1}{x^2} \cdot \left(y - x\frac{dy}{dx}\right)$$

$$= \frac{d}{dx}\left(\frac{y}{x}\right)$$

Solving the Linear First Order Differential Equation

For the general first order linear differential equation, we assume that an integrating factor, that is only a function of x, exists. Let us call it $\eta(x)$. Then this integrating factor, if multiplied by the expression in the differential equation should reduce the expression to an exact differential. Let us see how:

$$\frac{dy}{dx} + P(x)y = Q(x)$$

$$\eta(x)\frac{dy}{dx} + \eta(x)P(x)y = \eta(x)Q(x)$$

On inspection, note that if we put $\eta(x)P(x) = \eta'(x)$, we can easily get an exact differential on the left hand side by the:

Product Rule of Differentiation

$$\eta(x)\frac{dy}{dx} + \eta'(x)y = \eta(x)Q(x)$$

$$\frac{d}{dx}(\eta(x)y) = \eta(x)Q(x)$$

Now integrate on both sides with respect to x:

$$\int \frac{d}{dx}(\eta(x)y)dx = \int \eta(x)Q(x)dx$$

$$\eta(x)y + c = \int \eta(x)Q(x)dx$$

$$y = \frac{\int \eta(x)Q(x)dx - c}{\eta(x)}$$

Thus, we have got our general solution in the terms of our integrating function $\eta(x)$. You might worry that $\eta(x)$ is unknown to us. But on the contrary, it is very easy to compute in terms of other known quantities. If you remember, we had used the condition $\eta(x)P(x) = \eta'(x)$ to get the above solution. This is actually a variables separable equation which can be easily solved for $\eta(x)$ in the following way:

$$\eta(x)P(x) = \eta'(x)$$

$$P(x) = \frac{\eta'(x)}{\eta(x)}$$

$$\int P(x)dx = \int \frac{\eta'(x)}{\eta(x)}dx$$

$$\int P(x)dx = \ln(\eta(x)) + k$$

$$\eta(x) = e^{\int P(x)dx - k}$$

$$\eta(x) = Ke^{\int P(x)dx}, \text{ where } K = e^{-k}$$

This expresses the integrating function $\eta(x)$ in terms of an arbitrary constant K, and the known function $P(x)$. Let us now put this back into our solution of the differential equation:

$$y = \frac{\int \eta(x)Q(x)dx - c}{\eta(x)}$$

$$y = \frac{K\int e^{\int P(x)dx}Q(x)dx - c}{Ke^{\int P(x)dx}}$$

$$y = \frac{\int e^{\int P(x)dx}Q(x)dx - \frac{c}{K}}{e^{\int P(x)dx}}$$

Let us just denote the arbitrary constant $\frac{c}{K}$ as C which is simply another arbitrary constant. In a concise form, we can then write the differential equation along with its solution as:

$$\frac{dy}{dx} + P(x)y = Q(x)$$

has the solution, $y = \dfrac{\int \eta(x)Q(x)dx - C}{\eta(x)}$

where $\eta(x) = e^{\int P(x)dx}$

Thus, if we express the given differential equation in the standard form, we can now easily write down the solution using the above formula.

Example: Solve

$$x\frac{dy}{dx} - 2y = x^3 \cos 4x$$

Solution: The first step is to convert the given differential equation to the standard form $\frac{dy}{dx} + P(x)y = Q(x)$. On rearranging, we can obtain the given differential equation as:

$$\frac{dy}{dx} - \frac{2}{x}y = x^2\cos 4x$$

On comparing with the standard form, we get $P(x) = -\frac{2}{x}$ and $Q(x) = x^2\cos 4x$. We should now compute the integrating factor first, then proceed towards the general solution. From the formula $\eta(x) = e^{\int P(x)dx}$, we can work out the integrating factor here as:

$$\eta(x) = e^{\int\left(\frac{2}{x}\right)dx}$$
$$= e^{-2\ln x}$$
$$= e^{\ln(x^{-2})} = x^{-2}$$

Now we can use this in the formula that we had derived for the general solution.

$$y = \frac{\int \eta(x)Q(x)dx - C}{\eta(x)}$$

$$y = \frac{\int x^{-2}x^2\cos 4x\,dx - C}{x^2}$$

$$y = x^2(\frac{\sin 4x}{4} - C)$$

Example: Find the solution to the following differential equation.

$$\frac{dv}{dt} = 9.8 - 0.196v$$

Solution: First, we need to get the differential equation in the correct form.

$$\frac{dv}{dt} + 0.196v = 9.8$$

From this we can see that $p(t) = 0.196$ and so $\mu(t)$ is then.

$$\mu(t) = e^{\int 0.196dt} = e^{0.196t}$$

Officially there should be a constant of integration in the exponent from the integration. However, we can drop that for exactly the same reason that we dropped the k from.

Now multiply all the terms in the differential equation by the integrating factor and do some simplification.

$$e^{0.196t}\frac{dv}{dt} + 0.196e^{0.196t}v = 9.8\,e^{0.196t}$$

$$(e^{0.196t}v)' = 9.8\,e^{0.196t}.$$

Integrate both sides and don't forget the constants of integration that will arise from both integrals.

$$\int (e^{0.196t}v)' \, dt = \int 9.8e^{0.196t} \, dt$$

$$e^{0.196t}v + k = 50e^{0.196t} + c$$

Okay. It's time to play with constants again. We can subtract k from both sides to get.

$$e^{0.196t}v = 50e^{0.196t} + c - k$$

Both c and k are unknown constants and so the difference is also an unknown constant. We will therefore write the difference as c. So, we now have:

$$e^{0.196t}v = 50e^{0.196t} + c$$

From this point on we will only put one constant of integration down when we integrate both sides knowing that if we had written down one for each integral, as we should, the two would just end up getting absorbed into each other.

The final step in the solution process is then to divide both sides by $e^{0.196t}$ or to multiply both sides by $e^{-0.196t}$. Either will work, but we usually prefer the multiplication route. Doing this gives the general solution to the differential equation.

$$v(t) = 50 + ce^{-0.196t}$$

Example: Solve the following IVP.

$$\frac{dv}{dt} = 9.8 - 0.196v \quad v(0) = 48$$

Solution: To find the solution to an IVP we must first find the general solution to the differential equation and then use the initial condition to identify the exact solution that we are after. So, since this is the same differential equation as we looked at in Example above, we already have its general solution.

$$v = 50 + ce^{-0.196t}$$

Now, to find the solution we are after we need to identify the value of cc that will give us the solution we are after. To do this we simply plug in the initial condition which will give us an equation we can solve for cc. So, let's do this:

$$48 = v(0) = 50 + c \quad \Rightarrow \quad c = -2$$

So, the actual solution to the IVP is.

$$v = 50 - 2e^{-0.196t}$$

Example: Solve the following IVP.

$$\cos(x)y' + \sin(x)y = 2\cos^3(x)\sin(x) - 1 \quad y\left(\frac{\pi}{4}\right) = 3\sqrt{2}, \quad 0 \le x < \frac{\pi}{2}$$

Solution: Rewrite the differential equation to get the coefficient of the derivative a one.

$$y' + \frac{\sin(x)}{\cos(x)} y = 2\cos^2(x)\sin(x) - \frac{1}{\cos(x)}$$

$$y' + \tan(x)y = 2\cos^2(x)\sin(x) - \sec(x)$$

Now find the integrating factor:

$$\mu(t) = e^{\int \tan(x)dx} = e^{\ln|\sec(x)|} = e^{\ln \sec(x)} = \sec(x).$$

Can you do the integral? If not rewrite tangent back into sines and cosines and then use a simple substitution. We could drop the absolute value bars on the secant because of the limits on x. In fact, this is the reason for the limits on x. Note as well that there are two forms of the answer to this integral. They are equivalent as shown below. Which you use is really a matter of preference.

$$\int \tan(x)dx = -\ln|\cos(x)| = \ln|\cos(x)|^{-1} = \ln|\sec(x)|$$

We made use of the following fact,

$$e^{\ln f(x)} = f(x)$$

This is an important fact that you should always remember for these problems. We will want to simplify the integrating factor as much as possible in all cases and this fact will help with that simplification.

Now back to the example. Multiply the integrating factor through the differential equation and verify the left side is a product rule. Note as well that we multiply the integrating factor through the rewritten differential equation and NOT the original differential equation. Make sure that you do this. If you multiply the integrating factor through the original differential equation you will get the wrong solution.

$$\sec(x)y' + \sec(x)\tan(x)y = 2\sec(x)\cos^2(x)\sin(x) - \sec^2(x)$$

$$(\sec(x)y)' = 2\cos(x)\sin(x) - \sec^2(x)$$

Integrate both sides.

$$\int (\sec(x)y(x))' \, dx = \int 2\cos(x)\sin(x) - \sec^2(x)\, dx$$

$$\sec(x)y(x) = \int \sin(2x) - \sec^2(x)dx$$

$$\sec(x)\,y(x) = -\frac{1}{2}\cos(2x) - \tan(x) + c$$

The use of the trig formula $\sin(2\theta) = 2\sin\theta\cos\theta$ that made the integral easier. Next, solve for the solution.

$$y(x) = -\frac{1}{2}\cos(x)\cos(2x) - \cos(x)\tan(x) + c\cos(x)$$

$$= -\frac{1}{2}\cos(x)\cos(2x) - \sin(x) + c\cos(x)$$

Finally, apply the initial condition to find the value of c.

$$3\sqrt{2} = y\left(\frac{\pi}{4}\right) = -\frac{1}{2}\cos\left(\frac{\pi}{4}\right)\cos\left(\frac{\pi}{2}\right) - \sin\left(\frac{\pi}{4}\right) + c\cos\left(\frac{\pi}{4}\right)$$

$$3\sqrt{2} = -\frac{\sqrt{2}}{2} + c\frac{\sqrt{2}}{2}$$

$$c = 7$$

The solution is then.

$$y(x) = -\frac{1}{2}\cos(x)\cos(2x) - \sin(x) + 7\cos(x)$$

Below is a plot of the solution:

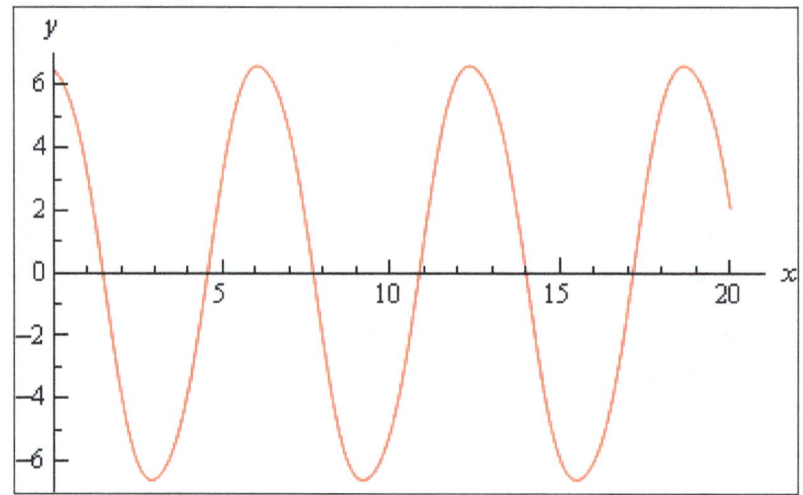

Example: Find the solution to the following IVP.

$$ty' + 2y = t^2 - t + 1 \quad y(1) = \frac{1}{2}$$

Solution: First, divide through by the t to get the differential equation into the correct form.

$$y' + \frac{2}{t}y = t - 1 + \frac{1}{t}$$

Now let's get the integrating factor, $\mu(t)$.

$$\mu(t) = e^{\int \frac{2}{t}dt} = e^{2\ln|t|}$$

Now, we need to simplify $\mu(t)$. However, we can't use equation $e^{\ln f(x)} = f(x)$ yet as that requires a coefficient of one in front of the logarithm. So, recall that:

$$\ln x^r = r \ln x$$

and rewrite the integrating factor in a form that will allow us to simplify it.

$$\mu(t) = e^{2\ln|t|} = e^{\ln|t|^2} = |t|^2 = t^2$$

We were able to drop the absolute value bars here because we were squaring the t, but often they can't be dropped so be careful with them and don't drop them unless you know that you can. Often the absolute value bars must remain.

Now, multiply the rewritten differential equation (remember we can't use the original differential equation here) by the integrating factor.

$$(t^2 y)' = t^3 - t^2 + t$$

Integrate both sides and solve for the solution.

$$t^2 y = \int t^3 - t^2 + t \, dt$$

$$= \frac{1}{4}t^4 - \frac{1}{3}t^3 + \frac{1}{2}t^2 + c$$

$$y(t) = \frac{1}{4}t^2 - \frac{1}{3}t + \frac{1}{2} + \frac{c}{t^2}$$

Finally, apply the initial condition to get the value of c.

$$\frac{1}{2} = y(1) = \frac{1}{4} - \frac{1}{3} + \frac{1}{2} + c \Rightarrow c = \frac{1}{12}$$

The solution is then,

$$y(t) = \frac{1}{4}t^2 - \frac{1}{3}t + \frac{1}{2} + \frac{1}{12t^2}$$

Here is a plot of the solution.

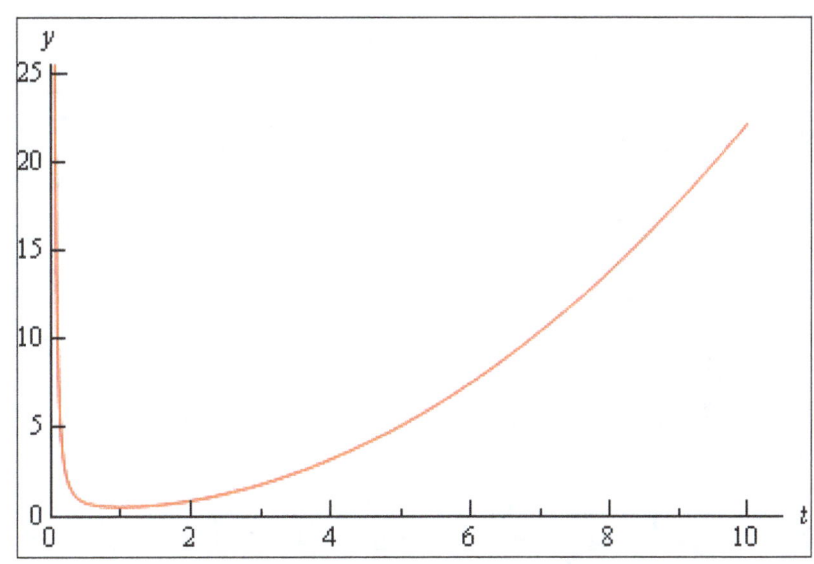

Example: Find the solution to the following IVP.

$$ty' - 2y = t^5 \sin(2t) - t^3 + 4t^4 \quad y(\pi) = \frac{3}{2}\pi^4$$

Solution: First, divide through by t to get the differential equation in the correct form.

$$y' - \frac{2}{t}y = t^4 \sin(2t) - t^2 + 4t^3$$

Now that we have done this we can find the integrating factor, $\mu(t)$.

$$\mu(t) = e^{\int -\frac{2}{t}dt} = e^{-2\ln|t|}$$

Do not forget that the "-" is part of p(t). Forgetting this minus sign can take a problem that is very easy to do and turn it into a very difficult, if not impossible problem so be careful.

Now, we just need to simplify this as we did in the previous example.

$$\mu(t) = e^{-2\ln|t|} = e^{\ln|t|-2} = |t|^{-2} = t^{-2}$$

Again, we can drop the absolute value bars since we are squaring the term.

Now multiply the differential equation by the integrating factor (again, make sure it's the rewritten one and not the original differential equation).

$$(t^{-2}y)' = t^2 \sin(2t) - 1 + 4t$$

Integrate both sides and solve for the solution.

$$t^{-2}y(t) = \int t^2 \sin(2t)dt + \int -1 + 4t \; dt$$

$$t^{-2}y(t) = -\frac{1}{2}t^2 \cos(2t) + \frac{1}{2}t\sin(2t) + \frac{1}{4}\cos(2t) - t + 2t^2 + c$$

$$y(t) = -\frac{1}{2}t^4 \cos(2t) + \frac{1}{2}t^3 \sin(2t) + \frac{1}{4}t^2 \cos(2t) - t^3 + 2t^4 + ct^2$$

Apply the initial condition to find the value of c.

$$\frac{3}{2}\pi^4 = y(\pi) = -\frac{1}{2}\pi^4 + \frac{1}{4}\pi^2 - \pi^3 + 2\pi^4 + c\pi^2 = \frac{3}{2}\pi^4 - \pi^3 + \frac{1}{4}\pi^2 + c\pi^2$$

$$\pi^3 - \frac{1}{4}\pi^2 = c\pi^2$$

$$c = \pi - \frac{1}{4}$$

The solution is then:

$$y(t) = -\frac{1}{2}t^4 \cos(2t) + \frac{1}{2}t^3 \sin(2t) + \frac{1}{4}t^2 \cos(2t) - t^3 + 2t^4 + \left(\pi - \frac{1}{4}\right)t^2$$

Below is a plot of the solution.

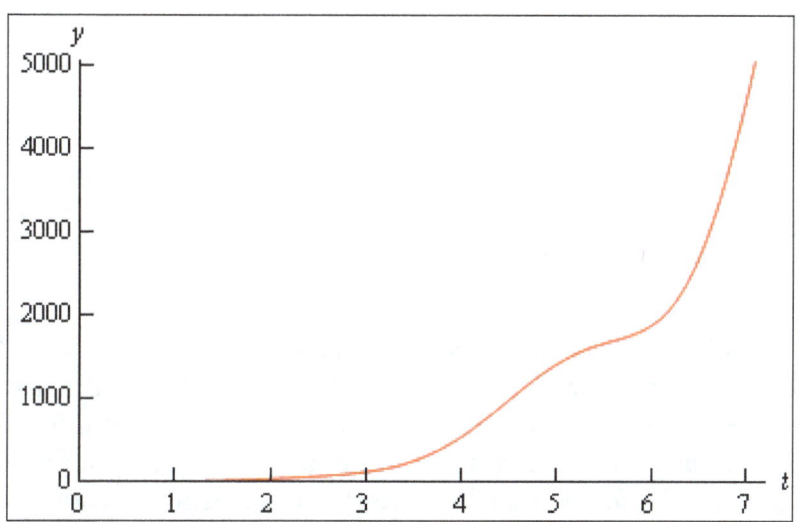

Example: Find the solution to the following IVP and determine all possible behaviors of the solution as t→∞. If this behavior depends on the value of y_0 give this dependence.

$$2y' - y = 4\sin(3t) \quad y(0) = y_0$$

Solution: First, divide through by a 2 to get the differential equation in the correct form.

$$y' - \frac{1}{2}y = 2\sin(3t)$$

Now find $\mu(t)$.

$$\mu(t) = e^{\int -\frac{1}{2}dt} = e^{-\frac{t}{2}}$$

Multiply $\mu(t)$ through the differential equation and rewrite the left side as a product rule.

$$\left(e^{-\frac{t}{2}}y\right)' = 2e^{-\frac{t}{2}}\sin(3t)$$

Integrate both sides (the right side requires integration by parts – you can do that right?) and solve for the solution.

$$e^{-\frac{t}{2}}y = \int 2e^{-\frac{t}{2}}\sin(3t)dt + c$$

$$e^{-\frac{t}{2}}y = -\frac{24}{37}e^{-\frac{t}{2}}\cos(3t) - \frac{4}{37}e^{-\frac{t}{2}}\sin(3t) + c$$

$$y(t) = -\frac{24}{37}\cos(3t) - \frac{4}{37}\sin(3t) + ce^{\frac{t}{2}}$$

Apply the initial condition to find the value of cc and note that it will contain y_0 as we don't have a value for that:

$$y_0 = y(0) = -\frac{24}{37} + c \Rightarrow c = y_0 + \frac{24}{37}$$

So, the solution is:

$$y(t) = -\frac{24}{37}\cos(3t) - \frac{4}{37}\sin(3t) + \left(y_0 + \frac{24}{37}\right)e^{\frac{t}{2}}$$

Now that we have the solution, let's look at the long term behavior (*i.e.* t→∞) of the solution. The first two terms of the solution will remain finite for all values of t. It is the last term that will determine the behavior of the solution. The exponential will always go to infinity as t→∞, however depending on the sign of the coefficient c (yes we've already found it, but for ease of this discussion we'll continue to call it c). The following table gives the long term behavior of the solution for all values of c.

Range of c	Behavior of solution as t→∞
$c < 0$	$y(t) \rightarrow -\infty$
$c = 0$	$y(t)$ remains finite
$c > 0$	$y(t) \rightarrow \infty$

This behavior can also be seen in the following graph of several of the solutions.

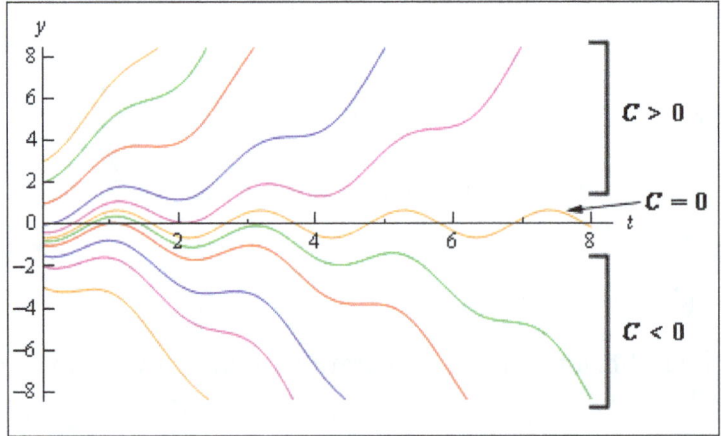

Now, because we know how c relates to y_0 we can relate the behavior of the solution to y_0. The following table gives the behavior of the solution in terms of y_0 instead of c.

Range of y_0	Behavior of solution as t→∞
$y_0 < -\dfrac{24}{37}$	$y(t) \rightarrow -\infty$

$y_0 = -\dfrac{24}{37}$	$y(t)$ remains finite
$y_0 > -\dfrac{24}{37}$	$y(t) \to \infty$

For $y_0 = -\dfrac{24}{37}$ the solution will remain finite. That will not always happen.

Bernoulli Differential Equation

The Bernoulli differential equation is an equation of the form $y' + p(x)y = q(x)y^n$. This is a non-linear differential equation that can be reduced to a linear one by a clever substitution. The new equation is a first order linear differential equation, and can be solved explicitly. The Bernoulli equation was one of the first differential equations to be solved, and is still one of very few non-linear differential equations that can be solved explicitly. Most other such equations either have no solutions, or solutions that cannot be written in a closed form, but the Bernoulli equation is an exception.

Example: The idea behind the Bernoulli equation is to substitute $v = y^{1-n}$, and work with the resulting equation, as shown in the example below.

Solve the differential equation $y' + y = xy^2$.

We start by dividing through by y^2 to get:

$$y' y^{-2} + y^{-1} = x.$$

Now, if we let $v = y^{-1}$, we have $v' = -y^{-2}y'$, so our equation becomes:

$$-v' + v = x \Rightarrow v' - v = -x$$

Now, we multiply through by the integrating factor e^{-x}, and we have:

$$e^{-x}v' - e^{-x}v = -xe^{-x} \Rightarrow \left[e^{-x}v\right]' = -xe^{-x} \Rightarrow e^{-x}v + e^{-x} + xe^{-x} + C \Rightarrow v = 1 + x + Ce^{x}.$$

Finally, since $v = y^{-1}$, we have the solution.

General Solution

The same procedure can be used to solve the general version of the equation.

Solve the differential equation $y' + p(x)y = q(x)y^n$.

We divide through by y^n, so we have:

$$y' y^{-n} + p(x)y^{1-n} = q(x).$$

Then, if $v = y^{1-n}$, $v' = (1-n)y^{-n}y'$, so this equation is:

$$\frac{1}{1-n}v' + p(x)v = q(x) \Rightarrow v' + (1-n)p(x)v = q(x)(1-n).$$

Then, the integrating factor will be a function $f(x)$ such that $f(x) = e^{\int (1-n)p(x)dx}$. We multiply through by this to get:

$$f(x)v' + p(x)(1-n)f(x)v = q(x)f(x)(1-n) \Rightarrow [f(x)v]' = q(x)f(x)(1-n) \Rightarrow v(x) = \frac{1-n}{f(x)}\int q(x)f(x)dx,$$

which is the solution to the Bernoulli equation.

Example: Solve the following IVP and find the interval of validity for the solution.

$$y' + \frac{4}{x}y = x^3y^2 \quad y(2) = -1, \quad x > 0$$

Solution: So, the first thing that we need to do is get this into the "proper" form and that means dividing everything by y^2. Doing this gives,

$$y^{-2}y' + \frac{4}{x}y^{-1} = x^3$$

The substitution and derivative that we'll need here is,

$$v = y^{-1} \quad v' = -y^{-2}y'$$

With this substitution the differential equation becomes,

$$-v' + \frac{4}{x}v = x^3$$

So, as noted above this is a linear differential equation that we know how to solve.

Here's the solution to this differential equation.

$$v' - \frac{4}{x}v = -x^3 \Rightarrow u(x) = e^{\int -\frac{4}{x}dx} = e^{-4\ln|x|} = x^{-4}$$

$$\int (x^{-4}v)'dx = \int -x^{-1}dx$$

$$x^{-4}v = -\ln|x| + c \Rightarrow v(x) = cx^4 - x^4\ln x$$

We dropped the absolute value bars on the x in the logarithm because of the assumption that x > 0.

Now we need to determine the constant of integration. This can be done in one of two ways. We can can convert the solution above into a solution in terms of y and then use the original initial condition or we can convert the initial condition to an initial condition in terms of v and use that. Because we'll need to convert the solution to y's eventually anyway and it won't add that much work in we'll do it that way.

So, to get the solution in terms of y all we need to do is plug the substitution back in. Doing this gives,

$$y^{-1} = x^4(c - \ln x)$$

At this point we can solve for y and then apply the initial condition or apply the initial condition and then solve for y. We'll generally do this with the later approach so let's apply the initial condition to get,

$$(-1)^{-1} = c2^4 - 2^4 \ln 2 \Rightarrow c = \ln 2 - \frac{1}{16}$$

Plugging in for cc and solving for y gives,

$$y(x) = \frac{1}{x^4\left(\ln 2 - \frac{1}{16} - \ln x\right)} = \frac{-16}{x^4(1 + 16\ln x - 16\ln 2)} = \frac{-16}{x^4\left(1 + 16\ln\frac{x}{2}\right)}$$

We did a little simplification in the solution. This will help with finding the interval of validity.

Before finding the interval of validity however, we could convert the original initial condition into an initial condition for v. Let's briefly talk about how to do that. To do that all we need to do is plug x=2 into the substitution and then use the original initial condition. Doing this gives,

$$v(2) = y^{-1}(2) = (-1)^{-1} = -1$$

So, in this case we got the same value for v that we had for y. Don't expect that to happen in general if you chose to do the problems in this manner.

Okay, let's now find the interval of validity for the solution. First, we already know that x>0 and that means we'll avoid the problems of having logarithms of negative numbers and division by zero at x=0. So, all that we need to worry about then is division by zero in the second term and this will happen where,

$$1 + 16\ln\frac{x}{2} = 0$$

$$\ln\frac{x}{2} = -\frac{1}{16}$$

$$\frac{x}{2} = e^{-\frac{1}{16}} \Rightarrow x = 2e^{-\frac{1}{16}} \approx 1.8788$$

The two possible intervals of validity are then,

$$0 < x < 2e^{-\frac{1}{16}} \quad 2e^{-\frac{1}{16}} < x < \infty$$

and since the second one contains the initial condition we know that the interval of validity is then $2e^{-\frac{1}{16}} < x < \infty$.

Here is a graph of the solution.

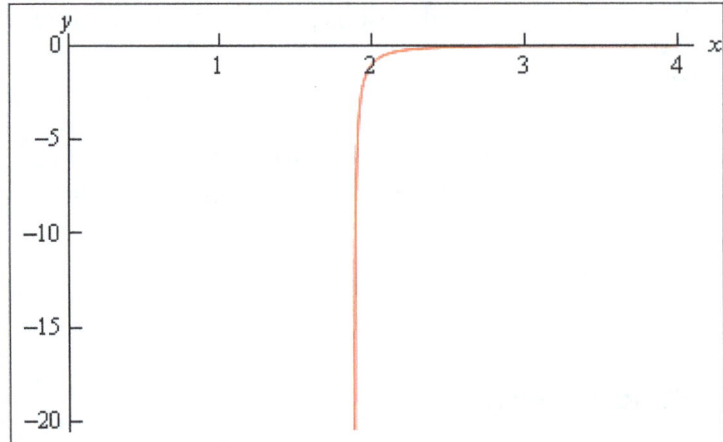

Example: Solve the following IVP and find the interval of validity for the solution.

$$y' = 5y + e^{-2x}y^{-2} \quad y(0) = 2$$

Solution: The first thing we'll need to do here is multiply through by y^2 and we'll also do a little rearranging to get things into the form we'll need for the linear differential equation. This gives,

$$y^2 y' - 5y^3 = e^{-2x}$$

The substitution here and its derivative is,

$$v = y^3 \quad v' = 3y^2 y'$$

Plugging the substitution into the differential equation gives,

$$\frac{1}{3}v' - 5v = e^{-2x} \quad \Rightarrow \quad v' - 15v = 3e^{-2x} \quad \mu(x) = e^{-15x}$$

We rearranged a little and gave the integrating factor for the linear differential equation solution. Upon solving we get,

$$v(x) = ce^{15x} - \frac{3}{17}e^{-2x}$$

Now go back to y's.

$$y^3 = ce^{15x} - \frac{3}{17}e^{-2x}$$

Applying the initial condition and solving for c gives,

$$8 = c - \frac{3}{17} \Rightarrow c = \frac{139}{17}$$

Plugging in c and solving for y gives,

$$y(x) = \left(\frac{139e^{15x} - 3e^{-2x}}{17} \right)^{\frac{1}{3}}$$

There are no problem values of x for this solution and so the interval of validity is all real numbers. Here's a graph of the solution.

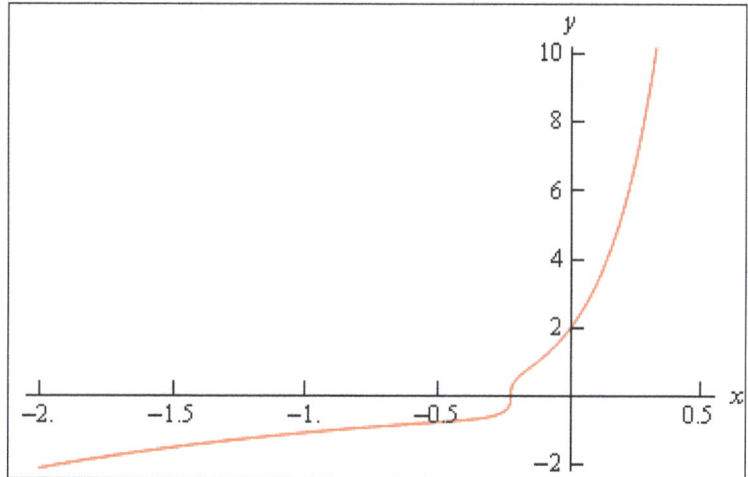

Example: Solve the following IVP and find the interval of validity for the solution.

$$6y' - 2y = xy^4 \qquad y(0) = -2$$

Solution: First get the differential equation in the proper form and then write down the substitution.

$$6y^{-4}y' - 2y^{-3} = x \implies v = y^{-3} \quad v' = -3y^{-4}y'$$

Plugging the substitution into the differential equation gives,

$$-2v' - 2v = x \implies v' + v = -\frac{1}{2}x \quad \mu(x) = e^x$$

Again, we've rearranged a little and given the integrating factor needed to solve the linear differential equation. Upon solving the linear differential equation we have,

$$v(x) = -\frac{1}{2}(x-1) + ce^{-x}$$

Now back substitute to get back into y's.

$$y^{-3} = -\frac{1}{2}(x-1) + ce^{-x}$$

Now we need to apply the initial condition and solve for c.

$$-\frac{1}{8} = \frac{1}{2} + c \implies c = -\frac{5}{8}$$

Plugging in cc and solving for y gives,

$$y(x) = -\frac{2}{(4x - 4 + 5e^{-x})\frac{1}{3}}$$

Next, we need to think about the interval of validity. In this case all we need to worry about it is division by zero issues and using some form of computational aid (such as Maple or Mathematica) we will see that the denominator of our solution is never zero and so this solution will be valid for all real numbers.

Here is a graph of the solution.

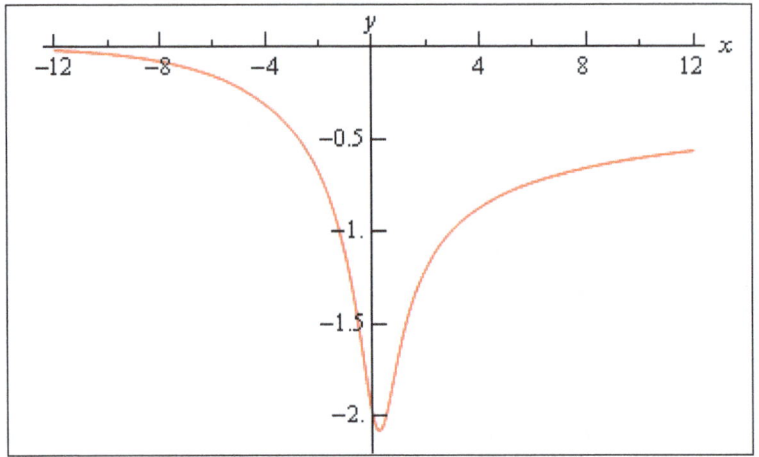

Example: Solve the following IVP and find the interval of validity for the solution.

$$y' + \frac{y}{x} - \sqrt{y} = 0 \quad y(1) = 0$$

Solution: Let's first get the differential equation into proper form.

$$y' + \frac{1}{x}y = y^{\frac{1}{2}} \quad \Rightarrow \quad y^{-\frac{1}{2}}y' + \frac{1}{x}y^{\frac{1}{2}} = 1$$

The substitution is then,

$$v = y^{\frac{1}{2}} \quad v' = \frac{1}{2}y^{-\frac{1}{2}}y'$$

Now plug the substitution into the differential equation to get,

$$2v' + \frac{1}{x}v = 1 \quad \Rightarrow \quad v' + \frac{1}{2x}v = \frac{1}{2} \quad \mu(x) = x^{\frac{1}{2}}$$

As we've done with the previous examples we've done some rearranging and given the integrating factor needed for solving the linear differential equation. Solving this gives us,

$$v(x) = \frac{1}{3}x + cx^{-\frac{1}{2}}$$

In terms of y this is,

$$y^{\frac{1}{2}} = \frac{1}{3}x + cx^{-\frac{1}{2}}$$

Applying the initial condition and solving for c gives,

$$0 = \frac{1}{3} + c \quad \Rightarrow \quad c = -\frac{1}{3}$$

Plugging in for c and solving for y gives us the solution.

$$y(x) = \left(\frac{1}{3}x - \frac{1}{3}x^{-\frac{1}{2}}\right)^2 = \frac{x^3 - 2x^{\frac{3}{2}} + 1}{9x}.$$

We multiplied everything out and converted all the negative exponents to positive exponents to make the interval of validity clear here. Because of the root (in the second term in the numerator) and the x in the denominator we can see that we need to require x > 0 in order for the solution to exist and it will exist for all positive x's and so this is also the interval of validity.

Here is the graph of the solution.

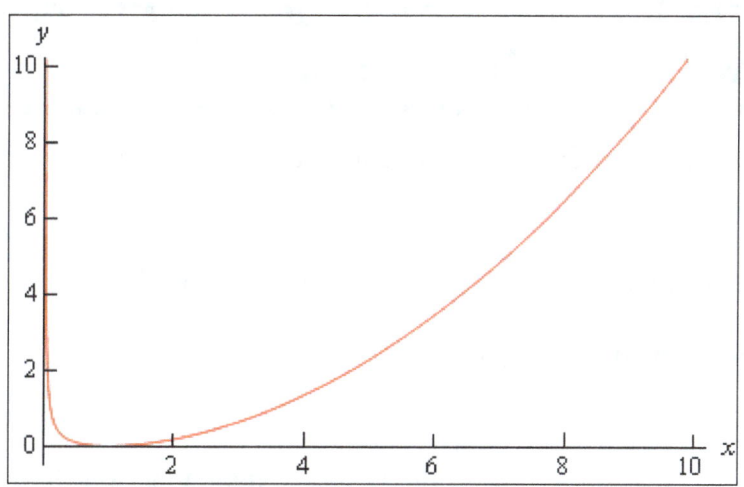

References

- Ordinary-differential-equation-introduction: mathinsight.org, Retrieved 16 July, 2019

- Homogeneous-differential-equations, differential-equations: toppr.com, Retrieved 17 January, 2019

- Linear-differential-equations, differential-equations: toppr.com, Retrieved 19 May, 2019

- Bernoullis-equation: brilliant.org, Retrieved 25 August, 2019

- Bernoulli: tutorial.math.lamar.edu, Retrieved 16 May, 2019

Linear Differential Equations of Second Order and Higher Order

3

Any differential equation which is defined by a linear polynomial in the unknown function is known as a linear differential equation. This chapter delves into the linear differential equations of the second order and higher order to provide a holistic understanding of differential equations. The topics elaborated in this chapter will help in gaining a better perspective about linear differential equations of second order and higher order.

Second Order Linear Homogeneous Differential Equations with Variable Coefficients

A linear homogeneous second order equation with variable coefficients can be written as:

$$y'' + a_1(x)y' + a_2(x)y = 0,$$

where $a_1(x)$ and $a_2(x)$ are continuous functions on the interval [a,b].

Linear Independence of Functions

Wronskian

The functions $y_1(x), y_2(x), \ldots, y_n(x)$ are called linearly dependent on the interval [a,b], if there are constants $\alpha_1, \alpha_2, \ldots, \alpha_n$, not all zero, such that for all values of x from this interval, the identity holds:

$$\alpha_1 y_1(x) + \alpha_2 y_2(x) + \ldots + \alpha_n y_n(x) \equiv 0$$

If this identity is satisfied only when $\alpha_1 = \alpha_2 = \ldots = \alpha_n = 0$, then these functions $y_1(x), y_2(x), \ldots, y_n(x)$ are called linearly independent on the interval [a,b].

For the case of two functions, the linear independence criterion can be written in a simpler form: The functions $y_1(x), y_2(x)$ are linearly independent on the interval [a,b], if their quotient in this segment is not identically equal to a constant:

$$\frac{y_1(x)}{y_2(x)} \neq const.$$

Otherwise, when $\frac{y_1(x)}{y_2(x)} \equiv const,$ these functions are linearly dependent.

Let n functions $y_1(x), y_2(x), \ldots, y_n(x)$ have derivatives of (n−1) order.

The determinant:

$$W(x) = W_{y_1, y_2, \ldots, y_n}(x) = \begin{vmatrix} y_1 & y_2 & \cdots & y_n \\ y'_1 & y'_2 & \cdots & y'_n \\ \cdots & \cdots & \cdots & \cdots \\ y_1^{(n-1)} & y_2^{(n-1)} & \cdots & y_n^{(n-1)} \end{vmatrix}$$

is called the Wronski determinant or Wronskian for this system of functions.

Wronskian Test

If the system of functions $y_1(x), y_2(x), \ldots, y_n(x)$ is linearly dependent on the interval [a,b], then its Wronskian vanishes on this interval.

It follows from here that if the Wronskian is nonzero at least at one point in the interval [a,b], then the functions $y_1(x), y_2(x), \ldots, y_n(x)$ are linearly independent. This property of the Wronskian allows to determine whether the solutions of a homogeneous differential equation are linearly independent.

Fundamental System of Solutions

A set of two linearly independent particular solutions of a linear homogeneous second order differential equation forms its fundamental system of solutions.

If $y_1(x), y_2(x)$ is a fundamental system of solutions, then the general solution of the second order equation is represented as:

$$y(x) = C_1 y_1(x) + C_2 y_2(x),$$

where C_1, C_2 are arbitrary constants.

For a given fundamental system of solutions $y_1(x)$, $y_2(x)$ we can construct the corresponding homogeneous differential equation. For the case of a second order equation, it is expressed in terms of the determinant:

$$\begin{vmatrix} y_1 & y_2 & y \\ y'_1 & y'_2 & y' \\ y''_1 & y''_2 & y'' \end{vmatrix} = 0.$$

Liouville's Formula

The general solution of a homogeneous second order differential equation is a linear combination of two linearly independent particular solutions $y_1(x)$, $y_2(x)$ of this equation.

Obviously, the particular solutions depend on the coefficients of the differential equation. The Liouville formula establishes a connection between the Wronskian W(x), constructed on the basis of particular solutions $y_1(x)$, $y_2(x)$, and the coefficient $a_1(x)$ in the differential equation.

Let W(x) be the Wronskian of the solutions $y_1(x)$, $y_2(x)$ of a linear second order homogeneous differential equation:

$$y'' + a_1(x)y' + a_2(x)y = 0,$$

in which the functions $a_1(x)$ and $a_2(x)$ are continuous on the interval [a,b]. Let the point x_0 belong to the interval [a,b]. Then for all $x \in$ [a, b] the Liouville formula:

$$W(x) = W(x_0)exp\left(-\int_{x_0}^{x} a_1(t)dt\right)$$

is valid.

Practical methods for solving second order homogeneous equations with variable coefficients:

Unfortunately, the general method of finding a particular solution does not exist. Usually this is done by guessing.

If a particular solution $y_1(x) \neq 0$ of the homogeneous linear second order equation is known, the original equation can be converted to a linear first order equation using the substitution $y = y_1(x)z(x)$ and the subsequent replacement z'(x)=u.

Another way to reduce the order is based on the Liouville formula. In this case, a particular solution $y_1(x)$ must also be known.

Example: Investigate whether the functions $y_1(x) = x + 2$, $y_2(x) = 2x - 1$ are linearly independent.

Solution: We form the quotient of two functions:

$$\frac{y_1(x)}{y_2(x)} = \frac{x+2}{2x-1} = \frac{x - \frac{1}{2} + \frac{5}{2}}{2x-1} = \frac{\frac{1}{2}(2x-1) + \frac{5}{2}}{2x-1} = \frac{1}{2} + \frac{5}{2(2x-1)}$$

$$= \frac{1}{2} + \frac{5}{4x-2}.$$

It is seen that this ratio is not equal to a constant, but depends on x. Hence, these functions are linearly independent.

Example: Find the Wronskian of the system of functions $y_1(x) = \cos x$, $y_2(x) = \sin x$.

Solution: The Wronskian of the system of two functions is calculated by the formula:

$$W_{y_1,y_2}(x) = \begin{vmatrix} y_1(x) & y_2(x) \\ y_1'(x) & y_2'(x) \end{vmatrix}.$$

Substituting the given functions and their derivatives, we obtain:

$$W_{y_1, y_2}(x) = \begin{vmatrix} \cos x & \sin x \\ -\sin x & \cos x \end{vmatrix} = \cos^2 x + \sin^2 x = 1.$$

It follows from here, that functions sin x and cos x are linearly independent.

Example: Write a homogeneous linear differential equation, if its fundamental system of solutions is known: x, e^x.

Solution: We find the derivatives of the given functions:

$$y_1' = x' = 1, y_1'' = 1' = 0;$$
$$y_2' = (e^x)' = e^x, y_2'' = (e^x)' = e^x.$$

The desired differential equation is expressed in terms of the determinant:

$$\begin{vmatrix} y_1 & y_2 & y \\ y_1' & y_2' & y' \\ y_1'' & y_2'' & y'' \end{vmatrix} = 0, \Rightarrow \begin{vmatrix} x & e^x & y \\ 1 & e^x & y' \\ 0 & e^x & y'' \end{vmatrix} = 0.$$

Expand the determinant along the 1st column:

$$x(e^x y'' - e^x y') - 1 \cdot (e^x y'' - e^x y) = 0, \Rightarrow e^x [(xy'' - xy') - (y'' - y)] = 0.$$

Given that $e^x \neq 0$, we obtain after simplification:

$$xy'' - xy' - y'' + y = 0, \Rightarrow (x-1)y'' - xy' + y = 0.$$

Example: Find the general solution of the equation $x^2 y'' - 2xy' + 2y = 0$, given the particular solution $y_1 = x$.

Solution: We make the substitution $y = y_1 z = xz$. Then,

$$y' = (xz)' = z + xz', y'' = (z + xz')' = z' + z' + xz'' = 2z' + xz''.$$

After substituting the original equation becomes:

$$x^2(2z' + xz'') - 2x(z + xz') + 2xz = 0, \Rightarrow 2x^2z' + x^3z'' - 2xz - 2x^2z'$$
$$+ 2xz = 0, \Rightarrow x^3 z'' = 0.$$

By replacing z'=p, we obtain a compact equation $x^3 p' = 0$, whose solution is the function p=C_1.

From here we find the function z:

$$p = C_1, \Rightarrow z' = C_1, \Rightarrow z = C_1 x + C_2.$$

Now we can write the general solution of the equation:

$$y(x) = xz = x(C_1 x + C_2) = C_1 x^2 + C_2 x.$$

Example: Find the general solution of the equation $(x + 1)y'' \quad 2y = 0$.

Solution: First, choose a particular solution for the given equation. Based on the structure of the equation, we can try to find a particular solution in the form of a quadratic function:

$$y_1 = Ax^2 + Bx + C.$$

Its derivatives will be equal to:

$$y_1' = 2Ax + B, y_1'' = 2A.$$

Substituting this into the differential equation, we determine the coefficients A, B, C:

$$(x^2 + 1) \cdot 2A - 2(Ax^2 + Bx + C) \equiv 0,$$
$$\Rightarrow 2Ax^2 + 2A - 2Ax^2 - 2Bx - 2C \equiv 0.$$

It is seen that the coefficients must satisfy the conditions:

$$\begin{cases} -2B = 0 \\ 2A - 2C = 0 \end{cases}, \Rightarrow \begin{cases} B = 0 \\ A = C \end{cases}.$$

A function of the form $y_1 = C(x^2 + 1)$ corresponds to these conditions. Thus, we can take the function $y_1 = x^2 + 1$ as a particular solution.

We make the following substitution to reduce the order of the equation:

$$y = y_1 z = (x^2 + 1)z, \Rightarrow y' = 2xz + (x^2 + 1)z',$$
$$\Rightarrow y'' = 2z + 2xz' + 2xz' + (x^2 + 1)z'' = 2z + 4xz' + (x^2 + 1)z''.$$

Then the equation becomes:

$$(x^2 + 1) \cdot [2z + 4xz' + (x^2 + 1)z''] - 2(x^2 + 1)z = 0, \Rightarrow (x^2 + 1) \cdot$$
$$[2z + 4xz' + (x^2 + 1)z'') - 2z] = 0, \Rightarrow (x^2 + 1)z'' + 4xz' = 0.$$

With another change z'=p(x), we obtain:

$$(x^2 + 1)p' + 4xp = 0.$$

Now we have a linear equation of the first order, which can be solved by separation of variables:

$$(x^2 + 1)\frac{dp}{dx} = -4xp, \Rightarrow \frac{dp}{p} = -\frac{4xdx}{x^2 + 1},$$
$$\Rightarrow \int \frac{dp}{p} = -2\int \frac{d(x^2 + 1)}{x^2 + 1}, \Rightarrow \ln|p| = -2\ln(x^2 + 1) + \ln C_1,$$
$$\Rightarrow \ln|p| = \ln \frac{C_1}{(x^2 + 1)^2}, \Rightarrow p = \frac{C_1}{(x^2 + 1)^2}.$$

We solve one more equation for z:

$$p = z' = \frac{C_1}{(x^2+1)^2}, \Rightarrow z = \int \frac{dx}{(x^2+1)^2}.$$

This integral is calculated using the reduction formula:

$$\int \frac{dx}{(x^2+1)^2} = \frac{x}{2(x^2+1)} + \frac{1}{2}\int \frac{dx}{x^2+1} = \frac{x}{2(x^2+1)} + \frac{1}{2}\arctan x + C_2.$$

As a result, we get the following expression for z:

$$z = C_1 \left[\frac{x}{2(x^2+1)} + \frac{1}{2}\arctan x + C_2 \right] = \frac{C_1 x}{x^2+1} + C_1 \arctan x + C_2,$$

where the constants C_1 and C_2 are renormalized to simplify the expression.

Thus, the general solution has the form:

$$y = (x^2+1)z = C_1 x + C_1(x^2+1)\arctan x + C_2(x^2+1)$$
$$= C_1[x + (x^2+1)\arctan x] + C_2(x^2+1).$$

Example: Find the general solution of the equation $x^2 y'' - 4xy' + 6y = 0$ using the Liouville formula. A particular solution of the equation is known and has the form: $y_1 = x^2$.

Solution: Let y_1 and y_2 be linearly independent particular solutions of the given equation (the solution y_1 is known). Then the Liouville formula is written as:

$$W(x) = W_{y_1, y_2}(x) = \begin{vmatrix} y_1 & y_2 \\ y_1' & y_2' \end{vmatrix} = W_0(x) exp\left(-\int_{x_0}^{x} \frac{a_1(t)}{a_0(t)} dt \right).$$

The integral in this expression is equal to:

$$\int_{x_0}^{x} \frac{a_1(t)}{a_0(t)} dt = \int_{x_0}^{x} \left(\frac{-4t}{t^2} \right) dt = (-4\ln t)\Big|_{x_0}^{x} = -4\ln x + 4\ln x_0$$

$$= -\ln x^4 + \ln x_0^4 = -\ln \frac{x^4}{x_0^4}.$$

Substituting this, we obtain:

$$\begin{vmatrix} y_1 & y_2 \\ y_1' & y_2' \end{vmatrix} = W_0(x) exp\left(\ln \frac{x^4}{x_0^4} \right) = \frac{W_0(x)}{x_0^4} x^4 = C_1 x^4,$$

where C_1 is an arbitrary constant.

As the particular solution y_1 is known, we get a first order differential equation for the determination of the other particular solution y_2. After dividing by y_1^2 we have:

$$y_2' y_1 - y_2 y_1' C_1 x^4 \Big| : y_1^2 \,, \Rightarrow \frac{y_2' y_1 - y_2 y_1'}{y_1^2} = \frac{C_1 x^4}{y_1^2}, \Rightarrow \left(\frac{y_2}{y_1}\right)' = \frac{C_1 x^4}{y_1^2}$$

$$= \frac{C_1 x^4}{x^4} = C_1.$$

From this we find the function y_2:

$$\frac{y_2}{y_1} = C_1 x + C_2, \Rightarrow y_2 = y_1(C_1 x + C_2) = x^2(C_1 x + C_2) = C_1 x^3 + C_2 x^2,$$

where C_1, C_2 are constants of integration. It is clear that y_2 is actually the general solution of the equation.

Example: Find the general solution of the equation $x^2 y'' + x y' - y = 0$ (for $x \neq 0$ by the Liouville formula, if a particular solution is known: $y_1 = x$.

Solution: Using the Liouville formula we get the following equation for the determination of the second particular solution y_2:

$$W(x) = W_{y_1, y_2}(x) = \begin{vmatrix} y_1 & y_2 \\ y_1' & y_2' \end{vmatrix} = C_1 \exp\left(-\int \frac{a_1(x)}{a_0(x)} dx\right).$$

The integral in this formula is:

$$\int \frac{a_1(x)}{a_0(x)} dx = \int \frac{x}{x^2} dx = \int \frac{dx}{x} = \ln|x|.$$

Then we can write:

$$y_2' y_1 - y_2 y_1' = C_1 e^{-\ln|x|}, \Rightarrow y_2' y_1 - y_2 y_1' = C_1 e^{\ln\frac{1}{|x|}}, \Rightarrow y_2' y_1 - y_2 y_1' = \frac{C_1}{x},$$

where C_1 is an arbitrary constant.

Divide both sides of the equation by $y_1^2 = x^2$ (assuming that $x \neq 0$).

$$\frac{y_2' y_1 - y_2 y_1'}{y_1^2} = \frac{C_1}{x y_1^2}, \Rightarrow \left(\frac{y_2}{y_1}\right)' = \frac{C_1}{x \cdot x^2} = \frac{C_1}{x^3}, \Rightarrow \frac{y_2}{y_1} = \int \frac{C_1}{x^3} dx$$

$$= -\frac{C_1}{2x^2} + C_2 = \frac{C_1}{x^2} + C_2.$$

Here, we redefined: $-\dfrac{C_1}{2} \to C_1$. The final answer is:

$$y_2 = y_1\left(\frac{C_1}{x^2} + C_2\right) = x\left(\frac{C_1}{x^2} + C_2\right) = \frac{C_1}{x} + C_2 x.$$

Second Order Linear Homogeneous Differential Equations with Constant Coefficients

The general second order homogeneous linear differential equation with constant coefficients is:

$$Ay'' + By' + Cy = 0,$$

where y is an unknown function of the variable x, and A, B, and C are constants. If A = 0 this becomes a first order linear equation, which in this case is separable, and so we already know how to solve. So we will consider the case $A \neq 0$. If we want, we can the divide through by A and obtain the equivalent equation:

$$y'' + by' + cy = 0,$$

where b B / A and $c = C / A$ (that is if we have nothing better to do, and like the first coefficient to be equal to 1).

Linear with constant coefficients means that each term in the Left Hand Side of the equation is a constant times y or a derivative of y. Homogeneous means that we exclude equations like $Ay'' + By' + Cy = f(x)$ which can be solved, in certain important cases, by an extension of the methods we will study here. Here we only will solve the case where the Right Hand Side f(x) is identically 0. Homogeneous also means that the constant function y = 0 is always a solution to the equation.

By now we know to expect 2 degrees of freedom in the solution of this second order equation, i.e., the general family of solutions should have two arbitrary constants. We call the general family of solutions for short the general solution. That means that to find our general solution we have to find two independent functions $y = f_1(x)$ and $y = f_2(x)$ which are solutions, and then the general solution will $y = C_1 \cdot f_1(x) + C_2 \cdot f_2(x)$. Now If $y = f_1(x)$ and $y = f_2(x)$ are indeed solutions, one can check by plugging in that $y = C_1 \cdot f_1(x) + C_2 \cdot f_2(x)$ will be a solution. The fact that all solutions are of this form (i.e., we haven't missed any solution) is harder to ascertain, but nevertheless true. Notice that of course we do need $f_1(x)$ and $f_2(x)$ to be independent. What does it mean for two functions to be independent? It means that one of them is not equal to a constant times the other. For example the functions $f_1(x) = e^x$ and $f_2(x) = 2e^x$ are not independent, because $f_2(x) = 2 \cdot f_1(x)$. On the other hand, for example the functions $g_1(x) = e^x$ and $g_2(x) = e^{2x}$ are independent (although we do not prove it, it seems intuitively clear that they are independent). To construct the general solution for a second order equation we do need two independent solutions. For example we cannot construct a general solution from say $f_1(x) = e^x$ and $f_2(x) = 2e^x$, because then the family $y = C_1 \cdot f_1(x) + C_2 \cdot f_2(x)$ will be $y = C_1 \cdot e^x + C_2 \cdot 2e^x = (C_1 + 2C_2)e^x = C_3 e^x$ (here $C_3 = C_1 + 2C_2$).

I.e., we only will get one degree of freedom if our functions are not independent, and we need two degrees of freedom. That is why it is important that we get two independent solutions of the equation somehow (by guessing, being smart or any way we can think of).

Let us look for the general solution of our second order homogeneous linear differential equation with constant coefficients. As often happens with differential equations, we try to make an educated guess as to the type of solutions. To do that one usually tries to see what happens with the equations with simplest possible coefficients. As we already agreed that we want $A \neq 0$, we can try checking what happens with the equation if we set the other coefficients to zero. And in fact so far we have already seen examples of 3 types of second order homogeneous linear differential equation with constant coefficients.

We have met in Calculus I with the equation (this is the simplest case when both b and c are equal to 0):

$$y'' = 0.$$

Here we don't need to guess, as we can solve by first getting $y' = C_2$, and then $y = C_1 + C_2 \cdot x$. Thus the general solution $y = C_1 + C_2 \cdot x$.

We can next try to solve the differential equation we get when $b = -1, c = 0$:

$$y'' - y' = 0, \quad y'' = y'$$

We try guessing among our usual collection of functions—constant $y = 1$, exponentials $y = e^{kx}$, trigonometry $y = \sin kx, y = \cos kx$—which one has the first derivative same as the second. We quickly come to the conclusion that of course the exponential $y = e^x$ (which has all its derivative equal among themselves), and the constant $y = 1$ both solve the equation. So the general solution is $y = C_1 \cdot e^x + C_2 \cdot 1$.

Next easiest in line to try to solve is the differential equation (with $b = 0, c = 1$):

$$y'' + y = 0, \quad y'' = -y.$$

Again we try to guess among our usual collection of functions which of them have second derivative equal to minus the function: that is the trigonometry $y = \sin x$ and $y = \cos x$. These two are indeed solutions, and they are independent, so our general solution is $y = C_1 \sin x + C_2 \cos x$.

In fact we saw another example of this situation:

$$4y'' + 25y = 0,$$

this was problem, and we saw that the solutions look like $y = C_1 \cdot \cos\frac{5}{2}x + C_2 \cdot \sin\frac{5}{2}x$. We have also seen:

$$2y'' + y' - y = 0,$$

this was problem, and we saw that the general solution was $y = C_1 \cdot e^{-x} + C_2 \cdot e^{\frac{1}{2}x}$.

Basically we see that the solutions in the different cases could be polynomials, they could be trigonometry, or they could be exponentials. Of course, by now we fully expect in more complicated cases the mixtures of the polynomial, trigonometry and exponential to show up: the solutions could possibly include functions of the kind of $x^n e^{rx} \cos kx$ and $x^n e^{rx} \sin kx$. In the case of second order the mixture of the three will not appear; only the mixture of polynomial-exponential and exponential-trigonometry. But in the higher order linear differential equations, one indeed meets with solutions containing the mixture of all three: polynomial-exponential-trigonometry. In this class though, we will only solve homogeneous linear differential equation of second order.

Let us try to find solution in the general case now; i.e., we start to look for solutions of the general equation of second order:

$$Ay'' + By' + Cy = 0,$$

Of course we always start with hoping for the best, so we try the easiest possible type of solution, which would be the exponential e^{rx} (easiest to differentiate). We plug in $y = e^{rx}$, trying to find what number r will make this a solution. That will give us:

$$Ar^2 e^{rx} + Bre^{rx} + Ce^{rx} = 0, \ (Ar^2 + Br + C)e^{rx} = 0.$$

But e^{rx} is never zero, so we can cancel it, and we get the equation:

$$Ar + Br + C$$

This is called the characteristic equation.

Characteristic Equation with different Roots

If the Discriminant of this quadratic equation $\left(D = B^2 - 4AC\right)$ is strictly positive this quadratic equation has two different solutions:

$$r_{1,2} = \frac{-B \pm \sqrt{D}}{2A},$$

thus we have at least two independent (different) choices for a solution of the differential equation: $e^{r_1 x}$ and $e^{r_2 x}$. And as by now we know to expect 2 degrees of freedom, this is perfect, because the general family of solutions will be:

$$y = C_1 \cdot e^{r_1 x} + C_2 \cdot e^{r_2 x}.$$

(It is easy to check by plugging in that this will be a solution of our differential equation for any C_1 and C_2. The fact that all solutions are of this form (i.e., we haven't missed any solution) is harder to ascertain, but nevertheless true).

Example: Solve the equation

$$y'' - y = 0,$$

Solution: We solve the characteristic equation ("replacing orders with powers", and using that y is the zero order derivative):

$$r^2 - 1 = 0, \ \text{which gives us } r_1 = 1; \ r_2 = -1$$

That means,

$$e^{1 \cdot x} = e^x \ \text{and} \ e^{-1 \cdot x} = e^{-x}$$

are two independent solutions. And thus the general solution is:

$$y = C_1 e^x + C_2 e^{-x}.$$

One can check that by plugging in the differential equation.

Example: Let's go back to the equation,

$$2y'' + y' - y = 0,$$

The characteristic equation is:

$$2r^2 + r - 1 = 0, \text{ which gives us } r_1 = -1; \; r_2 = \frac{1}{2},$$

which of course gives the solutions:

$$e^{-1 \cdot x} = e^{-x} \text{ and } e^{\frac{1}{2}x},$$

and thus the general solution is:

$$y = C_1 e^{-x} + C_2 e^{\frac{1}{2}x}.$$

One can check that by plugging in the differential equation.

Characteristic Equation with Repeated Roots

Let us now try and solve the equation:

$$y'' - 4y' + 4y = 0,$$

We write the characteristic equation:

$$r^2 - 4r + 4 = 0$$

and because the discriminant D = 16 − 16 = 0, we have $r_1 = r_2 = 2$. This only gives us the one solution e^{2x}, and no second solution for the second degree of freedom. How then to get the second solution? To get inspiration consider again the example:

$$y'' = 0,$$

Here we know the solution, y = $C_1 \cdot 1 + C_2 \cdot x$. But let us solve it with the characteristic equation:

r² = 0 which gives us both r_1 = 0 and r_2 = 0.

Now as we know the general solution: $y = C_1 \cdot 1 + C_2 \cdot x$, we can see that the first independent solution is indeed $e^{0 \cdot x} = 1$, but the second one appears to be x · $e^{0 \cdot x}$ = x. (Indeed x is a solution to the y" = 0—we can check by plugging in).

We can now go back to the equation $y'' - 4y' + 4y = 0$ and try something similar: the first independent solution is xe²ˣ, try with a second one: xe²ˣ . We check that it is a solution by plugging in:

$$\left(xe^{2x}\right)' = e^{2x} + 2xe^{2x}; \left(xe^{2x}\right)'' = 2e^{2x} + \left(2e^{2x} + 4xe^{2\xi}\right) = 4e^{2x} + 4xe^{2x}$$

Plugging in:

$$y'' - 4y' + 4y = \left(xe^{2x}\right)'' - 4\left(xe^{2x}\right)' + 4\left(xe^{2x}\right) = 4e^{2x} + 4xe^{2x} - 4\left(e^{2x} + 2xe^{2x}\right) + 4xe^{2x} = 0,$$

So xe^{2x} is indeed a solution of the differential equation $y'' - 4y' + 4y = 0.$ Then our general solution is $y = C_1 e^{2x} + C_2 xe^{2x}$.

We can do that every time we have repeated roots of the characteristic equation (i.e., when the discriminant $D = B^2 - 4AC = 0$):

If $r_1 = r_2 = r$ are the roots of the characteristic equation, then the first independent solution is $y = e^{rx}$, and the second one is $y = xe^{rx}$. The general solution is $y = C_1 e^{rx} + C_2 xe^{rx}$.

Example: Solve the equation,

$$y'' - 6y' + 9 = 0,$$

Solution: The characteristic equation is,

$$r^2 - 6r + 9 = 0, \ r_1 = r_2 = 3;$$

The two independent solutions are:

$$y_1 = e^{3x}, \ y_2 = xe^{3x};$$

the general solution is:

$$y = C_1 e^{3x} + C_2 xe^{3x}$$

One can check that by plugging in the differential equation.

Characteristic Equation with No Real Roots

Let us now try and solve again the following equation:

$$4y'' + 25y = 0,$$

but let's try through the characteristic equation. The problems is that the characteristic equation is

$$4r^2 + 25 = 0, \ \text{thus} \ r_{1,2} = \frac{-B \pm \sqrt{D}}{2A} = \frac{0 \pm \sqrt{-400}}{8} = \frac{\pm 5}{2}\sqrt{-1},$$

and it doesn't have real roots ($\sqrt{-1}$ is a slight problem). But we know that the solution of the differential equation is $y = C_1 \cdot \cos\frac{5}{2}x + C_2 \cdot \sin\frac{5}{2}x$. So we notice that the two independent solutions are the trigonometry $y = \cos kx$, and $y = \sin kx$, where $k = \frac{5}{2}$. We can get these if ignore the $\sqrt{-1}$ in $r_{1,2} = \frac{\pm 5}{2}\sqrt{-1}$, but instead of exponentials (which we would have if the $\sqrt{-1}$ was not there) we have trigonometry now.

Thus the conclusion seems to be that whenever we have real roots we take exponentials, whenever we have $\sqrt{-1}$ (imaginary roots) we take trigonometry.

Then the question is what happens with the mixtures? For example, let us consider the equation:

$$y'' + 4y' + 13y = 0,$$

The characteristic equation is:

$$r^2 - 4r + 13 = 0, \ r_{1,2} = \frac{-B \pm \sqrt{D}}{2A} = \frac{4 \pm \sqrt{-36}}{2} = 2 \pm 3\sqrt{-1}$$

Now we have a mixture: 2 is real (without $\sqrt{-1}$), and $3\sqrt{-1}$ is not. That suggests that we should have 2 appear in an exponential, and the 3 appear in a trigonometry (i.e., either cos 3x or sin 3x). Thus we concoct the mixtures $y = e^{2x} \cos 3x$ and $y = e^{2x} \cos 3x$ (indeed these are the only two independent mixtures possible in this case). We have to check if they are solutions: we can do that by plugging in. We calculate for $y = e^{2x} \cos 3x$:

$$y' = 2e^{2x} \cos 3x - 3e^{2x} \sin 3x$$
$$y' = 4e^{2x} \cos 3x - 6e^{2x} \sin 3x - 6e^{2x} \sin 3x - 9e^{2x} \cos 3x = -5^{2x} \cos 3x - 12e^{2x} \sin 3x$$

Plugging in:

$$y'' + 4y' + 13y = -5e^{2x} \cos 3x - 12e^{2x} \sin 3x - 4\left(2e^{2x} \cos 3x - 3e^{2x} \sin 3x\right) + 13e^{2x} \cos 3x =$$
$$= -5e^{2x} \cos 3x - 12e^{2x} \sin 3x - 8e^{2x} \cos 3x + 12e^{2x} \sin 3x + 13e^{2x} \cos 3x = 0.$$

So indeed $y = e^{2x} \cos 3x$ is a solution, and we can also check that $y = e^{2x} \sin 3x$ is a solution. Thus the general solution is:

$$y = C_1 \cdot e^{2x} \cos 3x + C_2 \cdot e^{2x} \sin 3x = e^{2x} \left(C_1 \cdot \cos 3x + C_2 \cdot \sin 3x\right).$$

So the rule then if the quadratic characteristic equation doesn't have real roots (the Discriminant $D = B^2 - 4AC < 0$).

Write the solutions of the characteristic equation carefully:

$$r_{1,2} = \frac{-B \pm \sqrt{D}}{2A} = \frac{-B}{2A} \pm \frac{\sqrt{D}}{2A} = \frac{-B}{2A} \pm (\frac{\sqrt{-D}}{2A})\sqrt{-1}$$

Then the two independent solutions are:

$$y_1 = e^{\frac{-B}{2A}x} \cos\left(\frac{\sqrt{D}}{2A}x\right), \ y_2 = e^{\frac{-B}{2A}x} \sin\left(\frac{\sqrt{D}}{2A}x\right);$$

and the general solution is:

$$y = C_1 y_1 + C_2 y_2 = C_1 e^{\frac{-B}{2A}x} \cos\left(\frac{\sqrt{-D}}{2A}x\right) + C_2 e^{\frac{-B}{2A}x} \sin\left(\frac{\sqrt{-D}}{2A}x\right)$$

$$= e^{\frac{-B}{2A}x}\left(C_1 \cos\left(\frac{\sqrt{-D}}{2A}x\right) + C_2 \sin\left(\frac{\sqrt{-D}}{2A}x\right)\right).$$

Example: Solve the equation,

$$y'' - 6y' + 10y = 0,$$

Solution: The characteristic equation is:

$$r - 6r + 10 = 0, \quad r_{1,2} = \frac{6 \pm \sqrt{-4}}{} = 3 \pm 1\sqrt{-1};$$

The two independent solutions are:

$$y_1 = e^{3x} \cos 1x = e^{3x} \cos x, \; y_2 = e^{3x} \sin x;$$

the general solution is:

$$y = C_1 e^{3x} \cos x + C_2 e^{3x} \sin x = e^{3x}\left(C_1 \cos x + C_2 \sin x\right).$$

Solving Initial Value Problems

Initial Value problem is a system of:

- Differential Equation.

- Initial Conditions.

Since a second order linear equation has two degrees of freedom, for an initial value problem we need two initial conditions: for y and for y'.

To solve the initial value problem:

- First: solve the differential equation. The general solution will have two constants c_1 and c_2 (because we have two degrees of freedom).

- Second: plug in the initial conditions and solve for the constants C_1 and C_2.

Example: Solve the initial value problem:

$$2y'' + y' - y = 0, \; y(0) = 0, \; y'(0) = 1$$

Solution:

- General solution: We already solved the equation, and we got the general solution,

$$y = C_1 e^{-x} + C_2 e^{\frac{1}{2}x}.$$

- Initial conditions: Now we plug in to find the constants C_1 and C_2,

$$y(0) = C_1 e - 0 + C_2 e^{\frac{1}{2} 0} = C_1 + C_2 = \text{should equal to } 0$$

$$y'(x) = -C_1 e^{-x} + C_2 \cdot \frac{1}{2} e^{\frac{1}{2} x}$$

$$y'(0) = -C_1 e^{-0} + C_2 \cdot \frac{1}{2} e^{\frac{1}{2} 0} = -C_1 + \frac{1}{2} \cdot C_2 = \text{should equal to } 1.$$

Thus we get the system for C_1 and C_2:

$$C_1 + C_2 = 0$$

$$-C_1 + \frac{1}{2} \cdot C_2 = 1$$

We solve this:

$$C_2 = -C_1$$

$$-C_1 - \frac{1}{2} \cdot C_1 = 1, -\frac{3}{2} \cdot C_1 = 1, C_1 = -\frac{2}{3}, C_2 = \frac{2}{3}.$$

Thus the solution to the Initial Value problem is:

$$y = -\frac{2}{3} e^{-x} + \frac{2}{3} e^{\frac{1}{2} x}$$

One can check that by plugging in both the differential equation and the initial conditions.

Example: Solve the initial value problem:

$$y'' - 4y' + 4y = 0, \quad y(0) = 1, \ y'(0) = 0$$

Solution:

- General solution: we already solved the equation, and we got the general solution,

$$y = C_1 e^{2x} + C_2 x e^{2x}.$$

- Initial conditions: Now we plug in to find the constants C_1 and C_2,

$$y(0) = C_1 e^0 + C_2 \cdot 0 \cdot e^0 = C_1 + 0 = \text{should equal to } 1$$

$$y'(x) = C_1 \cdot 2e^{2x} + C_2 \cdot \left(e^{2x} + 2x \cdot e^{2x}\right)$$

$$y'(0) = 2C_1 e^0 + C_2 \cdot \left(e^0 + 0 \cdot e^0\right) = 2C_1 + C_2 = \text{should equal to } 0.$$

Thus we get the system for C_1 and C_2:

$$C_1 = 1$$

$$2C_1 + C_2 = 1$$

We solve this:

$$C_1 = 1$$
$$C_2 = -2C_1 = -2, C_2 = -2.$$

Thus the solution to the Initial Value problem is:

$$y = e^{2x} - 2xe^{2x}.$$

One can check that by plugging in both the differential equation and the initial conditions.

Example: Solve the initial value problem,

$$y'' - 6y' + 10y = 0, \ y(0) = 1, \ y'(0) = 0$$

Solution:

- General solution: we already solved the equation, and we got the general solution

$$y = C_1 e^{3x} \cos x + C_2 e^{3x} \sin x = e^{3x} (C_1 \cos x + C_2 \sin x).$$

- Initial conditions: now we plug in to find the constants C_1 and C_2:

$$y(0) = C_1 e^0 \cos 0 + C_2 e^0 \sin 0 = C_1 + 0 = \text{should equal to } 1$$
$$y'(x) = (C_1 \cdot 3e^{3x} \cos x - C_1 \cdot e^{3x} \sin x) + (C_2 \cdot 3e^{3x} \sin x + C_2 e^{3x} \cos x)$$
$$y'(0) = (C_1 \cdot 3 \cdot 1 - C_1 \cdot 0) + (C_2 \cdot 3 \cdot 0 + C_2 \cdot 1 \cdot 1) = 3C_1 + C_2 = \text{should equal to } 0$$

Thus we get the system for C_1 and C_2:

$$C_1 = 1$$
$$3C_1 + C_2 = 1$$

We solve this:

$$C_1 = 1$$
$$C_2 = -3C_1 = -3, C_2 = -3$$

Thus the solution to the Initial Value problem is:

$$y = e^{3x} \cos x - 3e^{3x} \sin x = e^{3x} (\cos x - 3 \sin x).$$

One can check that by plugging in both the differential equation and the initial conditions.

Example: Solve the initial value problem,

$$y'' - 6y' + 10y = 0, \ y(0) = 1, y'(0) = 0$$

Solution:

- General solution: we already solved the equation, and we got the general solution,

$$y = C_1 e^{3x} \cos x + C_2 e^{3x} \sin x = e^{3x} \left(C_1 \cos x + C_2 \sin x \right).$$

- Initial conditions: now we plug in to find the constants C_1 and C_2,

$$y(0) = C_1 e^0 \cos^0 + C_2 e^0 \sin 0 = C_1 + 0 = \text{should equal to } 1$$

$$y'(x) = \left(C_1 \cdot 3e^{3x} \cos x - C_1 \cdot e^{3x} \sin x \right) + \left(C_2 \cdot 3e^{3x} \sin x + C_2 e^{3x} \cos x \right)$$

$$y'(0) = \left(C_1 \cdot 3 \cdot 1 - C_1 \cdot 0 \right) + \left(C_2 \cdot 3 \cdot 0 + C_2 \cdot 1 \cdot 1 \right) = 3C_1 + C_2 = \text{should equal to } 0.$$

Thus we get the system for C_1 and C_2:

$$C_1 = 1$$
$$3C_1 + C_2 = 1$$

We solve this:

$$C_1 = 1$$
$$C_2 = -3C_1 = -3, C_2 = -3.$$

Thus the solution to the Initial Value problem is:

$$y = e^{3x} \cos x - 3e^{3x} \sin x = e^{3x} \left(\cos x - 3 \sin x \right).$$

One can check that by plugging in both the differential equation and the initial conditions.

Connection between Exponentials and Trigonometry: Euler's Formula

One may ask how come in some of the cases the solutions to the differential equation are trigonometric functions, and in other cases the solutions are exponential functions; but in both cases the solutions are connected to the characteristic equation. The answer to that question is that there is a connection between the exponential $e^{(\lambda \cdot \sqrt{-1})x}$ and the trigonometric functions $\sin \lambda x$ and $\cos \lambda x$.

Of course to find that connection, one first need to make sense of what is $e^{(\lambda \sqrt{-1})x}$; i.e., one needs to work with numbers like $\lambda \cdot \sqrt{-1}$, which are not real numbers. That is achieved by giving the number $\sqrt{-1}$ formally a name i, so we write $i = \sqrt{-1}$; and we work with it as if nothing happened. We call this number i the imaginary unit (that's where the name i for the $\sqrt{-1}$ comes from). So we now work with numbers of the form $z = a + b \cdot i$, where the numbers a and b are any real numbers, and $i = \sqrt{-1}$. We can add and subtract such numbers as usual (as if the i is a variable, with of course the property $i^2 = (\sqrt{-1})^2 = -1$):

$$z_1 + z_2 = \left(a_1 + i \cdot b_1 \right) + \left(a_2 + i \cdot b_2 \right) = \left(a_1 + a_2 \right) + i \cdot \left(b_1 + b_2 \right),$$

$$z_1 - z_2 = \left(a_1 + i \cdot b_1 \right) - \left(a_2 + i \cdot b_2 \right) = \left(a_1 - a_2 \right) + i \cdot \left(b_1 - b_2 \right).$$

We can also multiply such numbers:

$$z_1 \cdot z_2 = \left(a_1 + i \cdot b_1 \right) \cdot \left(a_2 + i \cdot b_2 \right) = a_1 \cdot a_2 + i \cdot a_1 \cdot b_2 + i \cdot b_1 \cdot a_2 + i^2 \cdot b_1 \cdot b_2 =$$

$$= a_1 \cdot a_2 + i \cdot a_1 \cdot b_2 + i \cdot b_1 \cdot a_2 - b_1 \cdot b_2 = \left(a_1 \cdot a_2 - b_1 \cdot b_2 \right) + i \cdot \left(a_1 \cdot b_2 + b_1 \cdot a_2 \right).$$

Basically each time we just multiply out or add, but we keep the parts with i separate, and the part without i separate as well (of course we always keep in mind that $i^2 = (\sqrt{-1})^2 = -1$).

The part without i separate as well (of course we always keep in mind that $i^2 = (\sqrt{-1})^2 = -1$). The part without i is called real part, the part which goes with the i is called imaginary part. So for example the number 2 + 3i is a complex number, with real part equal to 2, and imaginary part equal to 3.

The collection of numbers $z = a + i \cdot b$ (or if one prefers to write it $a + b \cdot i$), where the numbers a and b are any real numbers, is called complex numbers.

So for example the number $2 + 3i$ is a complex number, with real part equal to 2, and imaginary part equal to 3.

The collection of numbers $z = a + i \cdot b$ (or if one prefers to write it $a + b \cdot i$), where the numbers a and b are any real numbers, is called complex numbers.

So with this newly obtained knowledge we can write the solutions of the characteristic equation when the discriminant is negative, i.e., D = B² − 4AC < 0, as:

$$r_{1,2} = \frac{-B \pm \sqrt{D}}{2A} = \frac{-B}{2A} \pm \frac{\sqrt{D}}{2A} = \frac{-B}{2A} \pm (\frac{\sqrt{-D}}{2A})\sqrt{-1} = \frac{-B}{2A} \pm (\frac{\sqrt{-D}}{2A})i.$$

So we see that when the discriminant is negative, the solutions are complex numbers, with the same real part $\frac{-B}{2A}$ and imaginary parts $\pm\frac{\sqrt{-D}}{2A}$. Call for short the real part $\mu = \frac{-B}{2A}$ and $\lambda = \frac{\sqrt{-D}}{2A}$. We want to understand what is $e^{\mu + \lambda i}$. Of course we know what is e^μ, when μ is real, but we don't know yet what is e^z, when z is complex. As we know how to add and multiply complex numbers we can use the Taylor expansion of the exponential to define what is e^z:

$$e^z = \sum_{n=0}^{\infty} \frac{z^n}{n!} = 1 + z + \frac{z^2}{2!} + \frac{z^3}{3!} + \frac{z^4}{4!} + \dots + \frac{z^n}{n!} + \dots$$

One can check that this exponential function is defined for all complex numbers, and it has all the usual properties the real exponential had, for example:

$$e^{z_1 + z_2} = e^{z_1} \cdot e^{z_2}.$$

Now if we put in the exponential $z = iy$, where y is real, we get:

$$e^{iy} = \sum_{n=0}^{\infty} \frac{(iy)^n}{n!} = 1 + iy + \frac{(iy)^2}{2!} + \frac{(iy)^3}{3!} + \frac{(iy)^4}{4!} + \frac{(iy)^5}{5!} + \dots + \frac{(iy)^n}{n!} + \dots =$$

$$= 1 + iy + \frac{i^2 y^2}{2!} + \frac{i^3 y^3}{3!} + \frac{i^4 y^4}{4!} + \frac{i^5 y^5}{5!} + \dots + \frac{i^n y^n}{n!} + \dots$$

Now we can use that:

$$i^2 = -1, \ i^3 = i^2 \cdot i = -i, \ i^4 = i^2 \cdot i^2 = 1, \ i^5 = i, \dots$$

That will mean that in the series formula for e^{iy} all the even powers will not have i, and all the odd powers will have i in them. I.e., the even powers will go into the real part of the e^{iy}, and the odd powers will go into the imaginary part of the e^{iy} :

$$e^{iy} = 1 + iy - \frac{y^2}{2!} - \frac{iy^3}{3!} + \frac{y^4}{4!} + \frac{iy^5}{5!} + \cdots$$

$$= (1 - \frac{y^2}{2!} + \frac{y^4}{4!} - \frac{x^6}{6!} + \ldots) + i(y\frac{y^3}{3!} + \frac{y^5}{5!} - \ldots)$$

Here we recognize that the first part, the real part, is the Taylor series of $\cos y$, and the second part, the imaginary part, is the Taylor series of $\sin y$. So we get the following famous result:

$$e^{iy} = \cos y + i\sin y.$$

This is called Euler's formula.

The Euler's formula explains what happens when the characteristic equation has complex roots: Suppose the roots of the characteristic equation are:

$$r_{1,2} = \frac{-B}{2A} \pm (\frac{\sqrt{-D}}{2A})i.$$

As before, call for short the real part $\mu = \frac{-B}{2A}$, and $\lambda = \frac{\sqrt{-D}}{2A}$, thus:

$$r_1 = \mu + i \cdot \lambda, \quad r_2 = \mu - i \cdot \lambda.$$

Then the general solution to the differential equation can be written using the complex exponential as:

$$y = C_1 e^{r_1 x} + C_2 e^{r_2 x} = C_1 e^{(\mu+i\cdot\lambda)x} + C_2 e^{(\mu-i\cdot\lambda)x} = C_1 e^{\mu x} e^{+i\cdot\lambda x} + C_2 e^{\mu x} e^{-i\cdot\lambda x} =$$
$$= C_1 e^{\mu x}\left(\cos \lambda x + i \cdot \sin \lambda x\right) + C_2 e^{\mu x}\left(\cos(-\lambda x) + i \cdot \sin(-\lambda x)\right)$$

Here we used the Euler's formula. Now we use the fact that $(-X) = \cos X$ and $\sin(-X) = -\sin X$ and we rearrange:

$$y = C_1 e^{\mu x}\left(\cos \lambda x + i\cdot\sin \lambda x\right) + C_2 e^{\mu x}\left(\cos \lambda x - i\cdot\sin \lambda x\right) = (C_1 + C_2)e^{\mu x} \cos \lambda x + i\cdot(C_1 - C_2)e^{\mu x}\sin \lambda x$$

Recall that we named $\mu = \frac{-B}{2A}$, and $\lambda = \frac{\sqrt{-D}}{2A}$, so we see that:

$$y = C_1 e^{r_1 x} + C_2 e^{r_2 x} = (C_1 + C_2)e^{\frac{-B}{2A}x}\cos(\frac{\sqrt{-D}}{2A}x) + i \cdot (C_1 - C_2)e^{\frac{-B}{2A}x}\sin(\frac{\sqrt{-D}}{2A})x.$$

Now if we call $\tilde{C}_1 = (C_1 + C_2)$ and $\tilde{C}_2 = i(C_1 - C_2)$, we get:

$$y = C_1 e^{r_1 x} + C_2 e^{r_2 x} = \tilde{C}_1 e^{\frac{-B}{2A}x}\cos(\frac{\sqrt{-D}}{2A}x) + \tilde{C}_2 e^{\frac{-B}{2A}x}\sin(\frac{\sqrt{-D}}{2A})x.$$

That shows that the general solution of the second order differential equation can be written in two equally correct ways: either we use the two complex exponentials, or the two exponential/trigonometry mixtures. The complex exponentials have the benefit of having the same form as the solutions for real characteristic roots. The mixtures though have the benefit of giving real solutions. Usually when solving the differential equation, or an initial value problem, we need a real solution, so the exponential/trigonometry mixtures are to be preferred.

Second Order Linear Non-homogeneous Differential Equations

Consider the nonhomogeneous linear differential equation:

$$a_2(x)y'' + a_1(x)y' + a_0(x)y = r(x).$$

The associated homogeneous equation:

$$a_2(x)y'' + a_1(x)y' + a_0(x)y = 0$$

is called the complementary equation. We will see that solving the complementary equation is an important step in solving a nonhomogeneous differential equation.

Particular Solution

A solution $y_p(x)$ of a differential equation that contains no arbitrary constants is called a *particular solution* to the equation.

General Solution to a Nonhomogeneous Equation

Let $y_p(x)$ be any particular solution to the nonhomogeneous linear differential equation:

$$a_2(x)y'' + a_1(x)y' + a_0(x)y = r(x).$$

Also, let $c_1 y_1(x) + c_2 y_2(x)$ denote the general solution to the complementary equation. Then, the general solution to the nonhomogeneous equation is given by:

$$y(x) = c_1 y_1(x) + c_2 y_2(x) + y_p(x).$$

Proof:

To prove $y(x)$ is the general solution, we must first show that it solves the differential equation and, second, that any solution to the differential equation can be written in that form. Substituting $y(x)$ into the differential equation, we have:

$$a_2(x)y'' + a_1(x)y' + a_0(x)y = a_2(x)(c_1 y_1 + c_2 y_2 + y_p)'' + a_1(x)(c_1 y_1 + c_2 y_2 + y_p)'$$
$$+ a_0(x)(c_1 y_1 + c_2 y_2 + y_p)$$

$$= [a_2(x)(c_1y_1 + c_2y_2)'' + a_1(x)(c_1y_1 + c_2y_2)' + a_0(x)(c_1y_1 + c_2y_2)]$$
$$+ a_2(x)y_p'' + a_1(x)y_p' + a_0(x)y_p$$
$$= 0 + r(x)$$
$$= r(x).$$

So $y(x)$ is a solution.

Now, let $z(x)$ be any solution to $a_2(x)y'' + a_1(x)y' + a_0(x)y = r(x)$. Then:

$$a_2(x)(z - yp)'' + a_1(x)(z - yp)' + a_0(x)(z - yp) = (a_2(x)z'' + a_1(x)z' + a_0(x)z)$$

$$-(a_2(x)yp'' + a_1(x)yp' + a_0(x)yp)$$

$$= r(x) - r(x)$$

$$= 0,$$

so $z(x) - y_p(x)$ is a solution to the complementary equation. But, $c_1y_1(x) + c_2y_2(x)$ is the general solution to the complementary equation, so there are constants c_1 and c_2 such that:

$$z(x) - yp(x) = c_1y_1(x) + c_2y_2(x).$$

Hence, we see that:

$$z(x) = c_1y_1(x) + c_2y_2(x) + y_p(x).$$

Example: Verifying the General Solution.

Given that $y_p(x) = x$ is a particular solution to the differential equation $y'' + y = x$, write the general solution and check by verifying that the solution satisfies the equation.

Solution: The complementary equation is $y'' + y = 0$, which has the general solution $c_1 \cos x + c_2 \sin x$. So, the general solution to the nonhomogeneous equation is:

$$y(x) = c_1 \cos x + c_2 \sin x + x.$$

To verify that this is a solution, substitute it into the differential equation. We have:

$$y'(x) = -c_1 \sin x + c_2 \cos x + 1$$

and

$$y''(x) = -c_1 \cos x - c_2 \sin x.$$

Then

$$y''(x) + y(x) = -c_1 \cos x - c_2 \sin x + c_1 \cos x + c_2 \sin x + x$$
$$= x.$$

So, $y(x)$ is a solution to $y'' + y = x$.

We learned how to solve homogeneous equations with constant coefficients. Therefore, for non-homogeneous equations of the form $ay''+by'+cy=r(x)$, we already know how to solve the complementary equation, and the problem boils down to finding a particular solution for the non-homogeneous equation. We now examine two techniques for this: the method of undetermined coefficients and the method of variation of parameters.

Undetermined Coefficients

The method of *undetermined coefficients* involves making educated guesses about the form of the particular solution based on the form of $r(x)$. When we take derivatives of polynomials, exponential functions, sines, and cosines, we get polynomials, exponential functions, sines, and cosines. So when $r(x)$ has one of these forms, it is possible that the solution to the nonhomogeneous differential equation might take that same form. Let's look at some examples to see how this works.

Example: Undetermined Coefficients when r(x) is a Polynomial.

Find the general solution to $y''+4y'+3y=3x$.

Solution: The complementary equation is $y''+4y'+3y=0$, with general solution $c_1 e-x+c_2 e^{-3x}$. Since $r(x)=3x$, the particular solution might have the form $y_p(x)=Ax+B$. If this is the case, then we have $y_p'(x)=A$ and $y_p''(x)=0$. For y_p to be a solution to the differential equation, we must find values for A and B such that:

$$y''+4y'+3y=3x$$
$$0+4(A)+3(Ax+B)=3x$$
$$3Ax+(4A+3B)=3x.$$

Setting coefficients of like terms equal, we have:

$$3A=3$$
$$4A+3B=0.$$

Then, $A=1$ and $B=-\dfrac{4}{3}$, so $y_p(x)=x-\dfrac{4}{3}$ and the general solution is:

$$y(x)=c_1 e^{-x}+c_2 e^{-3x}+x-\dfrac{4}{3}.$$

In example above notice that even though r(x) did not include a constant term, it was necessary for us to include the constant term in our guess. If we had assumed a solution of the form $yp=Ax$ (with no constant term) we would not have been able to find a solution. If the function r(x) is a polynomial, our guess for the particular solution should be a polynomial of the same degree, and it must include all lower-order terms, regardless of whether they are present in r(x).

Example: Determine a particular solution to,

$$y''-4y'-12y=3e^{5t}.$$

Solution: The point here is to find a particular solution, however the first thing that we're going to do is find the complementary solution to this differential equation. Recall that the complementary solution comes from solving,

$$y'' - 4y' - 12y = 0$$

The characteristic equation for this differential equation and its roots are:

$$r^2 - 4r - 12 = (r-6)(r+2) = 0 \implies r_1 = -2, r_2 = 6$$

The complementary solution is then,

$$y_c(t) = c_1 e^{-2t} + c_2 e^{6t}$$

At this point the reason for doing this first will not be apparent, however we want you in the habit of finding it before we start the work to find a particular solution. Eventually, as we'll see, having the complementary solution in hand will be helpful and so it's best to be in the habit of finding it first prior to doing the work for undetermined coefficients.

Now, let's proceed with finding a particular solution. As mentioned prior to the start of this example we need to make a guess as to the form of a particular solution to this differential equation. Since g(t) is an exponential and we know that exponentials never just appear or disappear in the differentiation process it seems that a likely form of the particular solution would be:

$$Y_P(t) = Ae^{5t}$$

Now, all that we need to do is do a couple of derivatives, plug this into the differential equation and see if we can determine what A needs to be.

Plugging into the differential equation gives:

$$25Ae^{5t} - 4(5Ae^{5t}) - 12(Ae^{5t}) = 3e^{5t}$$
$$-7Ae^{5t} = 3e^{5t}$$

So, in order for our guess to be a solution we will need to choose A so that the coefficients of the exponentials on either side of the equal sign are the same. In other words we need to choose A so that,

$$-7A = 3 \implies A = -\frac{3}{7}$$

Okay, we found a value for the coefficient. This means that we guessed correctly. A particular solution to the differential equation is then,

$$Y_P(t) = -\frac{3}{7}e^{5t}$$

Example: Solve the following IVP,

$$y'' - 4y' - 12y = 3e^{5t} \qquad y(0) = \frac{18}{7} \quad y'(0) = -\frac{1}{7}$$

Solution: We know that the general solution will be of the form,

$$y(t) = y_c(t) + Y_p(t)$$

and we already have both the complementary and particular solution from the first example so we don't really need to do any extra work for this problem.

One of the more common mistakes in these problems is to find the complementary solution and then, because we're probably in the habit of doing it, apply the initial conditions to the complementary solution to find the constants. This however, is incorrect. The complementary solution is only the solution to the homogeneous differential equation and we are after a solution to the non-homogeneous differential equation and the initial conditions must satisfy that solution instead of the complementary solution.

So, we need the general solution to the nonhomogeneous differential equation. Taking the complementary solution and the particular solution that we found in the previous example we get the following for a general solution and its derivative:

$$\left(D^4 + 5D^2 + 4\right)v = 4$$

Now, apply the initial conditions to these:

$$\frac{18}{7} = y(0) = c_1 + c_2 - \frac{3}{7}$$

$$-\frac{1}{7} = y'(0) = -2c_1 + 6c_2 - \frac{15}{7}$$

Solving this system gives $c_1 = 2$ and $c_2 = 1$. The actual solution is then:

$$y(t) = 2e^{-2t} + e^{6t} - \frac{3}{7}e^{5t}$$

Example: Find a particular solution for the following differential equation,

$$y'' - 4y' - 12y = \sin(2t)$$

Solution: Again, let's note that we should probably find the complementary solution before we proceed onto the guess for a particular solution. However, because the homogeneous differential equation for this example is the same as that for the first example we won't bother with that here.

Now, let's take our experience from the first example and apply that here. The first example had an exponential function in the g(t) and our guess was an exponential. This differential equation has a sine so let's try the following guess for the particular solution:

$$Y_p(t) = A\sin(2t)$$

Differentiating and plugging into the differential equation gives:

$$-4A\sin(2t) - 4(2A\cos(2t)) - 12(A\sin(2t)) = \sin(2t)$$

Collecting like terms yields:

$$-16A\sin(2t) - 8A\cos(2t) = \sin(2t)$$

We need to pick A so that we get the same function on both sides of the equal sign. This means that the coefficients of the sines and cosines must be equal. Or,

$$\cos(2t) : -8A = 0 \Rightarrow A = 0$$

$$\sin(2t) : -16A = 1 \Rightarrow A = -\frac{1}{16}$$

Notice two things. First, since there is no cosine on the right hand side this means that the coefficient must be zero on that side. More importantly we have a serious problem here. In order for the cosine to drop out, as it must in order for the guess to satisfy the differential equation, we need to set $A = 0$, but if $A = 0$, the sine will also drop out and that can't happen. Likewise, choosing A to keep the sine around will also keep the cosine around.

What this means is that our initial guess was wrong. If we get multiple values of the same constant or are unable to find the value of a constant then we have guessed wrong.

One of the nicer aspects of this method is that when we guess wrong our work will often suggest a fix. In this case the problem was the cosine that cropped up. So, to counter this let's add a cosine to our guess. Our new guess is:

$$Y_p(t) = A\cos(2t) + B\sin(2t)$$

Plugging this into the differential equation and collecting like terms gives,

$$-4A\cos(2t) - 4B\sin(2t) - 4(-2A\sin(2t) + 2B\cos(2t)) -$$
$$12(A\cos(2t) + B\sin(2t)) = \sin(2t)$$
$$(-4A - 8B - 12A)\cos(2t) + (-4B + 8A - 12B)\sin(2t) = \sin(2t)$$
$$(-16A - 8B)\cos(2t) + (8A - 16B)\sin(2t) = \sin(2t)$$

Now, set the coefficients equal:

$$\cos(2t): \quad -16A - 8B = 0$$
$$\sin(2t): \quad 8A - 16B = 1$$

Solving this system gives us:

$$A = \frac{1}{40} \qquad B = -\frac{1}{20}$$

We found constants and this time we guessed correctly. A particular solution to the differential equation is then,

$$Y_p(t) = \frac{1}{40}\cos(2t) - \frac{1}{20}\sin(2t)$$

Example: Find a particular solution for the following differential equation,

$$y'' - 4y' - 12y = 2t^3 - t + 3$$

Solution: Once, again we will generally want the complementary solution in hand first, but again we're working with the same homogeneous differential equation (you'll eventually see why we keep working with the same homogeneous problem) so we'll again just refer to the first example.

For this example, g(t) is a cubic polynomial. For this we will need the following guess for the particular solution.

$$Y^P(t) = At^3 + Bt^2 + Ct + D$$

Notice that even though g(t) doesn't have a t² in it our guess will still need one. So, differentiate and plug into the differential equation.

$$6At + 2B - 4(3At^2 + 2Bt + C) - 12(At^3 + Bt^2 + Ct + D) = 2t^3 - t + 3$$
$$-12At^3 + (-12A - 12B)t^2 + (6A - 8B - 12C)t + 2B - 4C - 12D = 2t^3 - t + 3$$

Now, as we've done in the previous examples we will need the coefficients of the terms on both sides of the equal sign to be the same so set coefficients equal and solve:

$$t^3: \qquad -12A = 2 \qquad \Rightarrow A = -\frac{1}{6}$$

$$t^2: \qquad -12A - 12B = 0 \qquad \Rightarrow B = \frac{1}{6}$$

$$t^1: \quad 6A - 8B - 12C = -1 \qquad \Rightarrow C = -\frac{1}{9}$$

$$t^0: \quad 2B - 4C - 12D = 3 \qquad \Rightarrow D = -\frac{5}{27}$$

Notice that in this case it was very easy to solve for the constants. The first equation gave A. Then once we knew A the second equation gave B, *etc.* A particular solution for this differential equation is then:

$$Y_P(t) = -\frac{1}{6}t^3 + \frac{1}{6}t^2 - \frac{1}{9}t - \frac{5}{27}$$

Example: Find a particular solution for the following differential equation.

$$y'' - 4y' - 12y = te^{4t}$$

Solution: You're probably getting tired of the opening comment, but again finding the complementary solution first really a good idea but again we've already done the work in the first example so we won't do it again here. We promise that eventually you'll see why we keep using the same homogeneous problem and why we say it's a good idea to have the complementary solution in hand first. At this point all we're trying to do is reinforce the habit of finding the complementary solution first.

Okay, let's start off by writing down the guesses for the individual pieces of the function. The guess for the t would be:

$$At + B$$

while the guess for the exponential would be:

$$Ce^{4t}$$

Now, since we've got a product of two functions it seems like taking a product of the guesses for the individual pieces might work. Doing this would give:

$$Ce^{4t}(At + B)$$

However, we will have problems with this. As we will see, when we plug our guess into the differential equation we will only get two equations out of this. The problem is that with this guess we've got three unknown constants. With only two equations we won't be able to solve for all the constants.

This is easy to fix however. Let's notice that we could do the following:

$$Ce^{4t}(At + B) = e^{4t}(ACt + BC)$$

If we multiply the C through, we can see that the guess can be written in such a way that there are really only two constants. So, we will use the following for our guess:

$$Y_P(t) = e^{4t}(At + B)$$

This is nothing more than the guess for the tt with an exponential tacked on for good measure.

Now that we've got our guess, let's differentiate, plug into the differential equation and collect like terms:

$$e^{4t}(16At + 16B + 8A) - 4(e^{4t}(4At + 4B + A)) - 12(e^{4t}(At + B)) = te^{4t}$$
$$(16A - 16A - 12A)te^{4t} + (16B + 8A - 16B - 4A - 12B)e^{4t} = te^{4t}$$
$$-12Ate^{4t} + (4A - 12B)e^{4t} = te^{4t}$$

When we're collecting like terms we want the coefficient of each term to have only constants in it. Following this rule we will get two terms when we collect like terms. Now, set coefficients equal:

$$te^{4t}: \qquad -12A = 1 \qquad \Rightarrow A = -\frac{1}{12}$$

$$e^{4t}: \qquad 4A - 12B = 0 \qquad \Rightarrow B = -\frac{1}{36}$$

A particular solution for this differential equation is then:

$$Y_P(t) = e^{4t}\left(-\frac{t}{12} - \frac{1}{36}\right) = -\frac{1}{36}(3t + 1)e^{4t}$$

Example: Write down the form of the particular solution to,

$$y'' + p(t)y' + q(t)y = g(t)$$

$$g(t) = 16e^{7t}\sin(10t)$$

$$g(t) = (9t^2 - 103t)\cos t$$

$$g(t) = -e^{-2t}(3 - 5t)\cos(9t)$$

Solution:

$$g(t) = 16e^{7t}\sin(10t)$$

So, we have an exponential in the function. Remember the rule. We will ignore the exponential and write down a guess for $16\sin(10t)$ then put the exponential back in.

The guess for the sine is:

$$A\cos(10t) + B\sin(10t)$$

Now, for the actual guess for the particular solution we'll take the above guess and tack an exponential onto it. This gives:

$$Y_p(t) = e^{7t}(A\cos(10t) + B\sin(10t))$$

One final note before we move onto the next part. The 16 in front of the function has absolutely no bearing on our guess. Any constants multiplying the whole function are ignored.

$$g(t) = (9t^2 - 103t)\cos t$$

We will start this one the same way that we initially started the previous example. The guess for the polynomial is:

$$At^2 + Bt + C$$

and the guess for the cosine is:

$$D\cos t + E\sin t$$

If we multiply the two guesses we get:

$$(At^2 + Bt + C)(D\cos t + E\sin t)$$

Let's simplify things up a little. First multiply the polynomial through as follows:

$$(At^2 + Bt + C)(D\cos t) + (At^2 + Bt + C)(E\sin t)$$

$$(ADt^2 + BDt + CD)\cos t + (AEt^2 + BEt + CE)\sin t$$

Notice that everywhere one of the unknown constants occurs it is in a product of unknown constants. This means that if we went through and used this as our guess the system of equations that we would need to solve for the unknown constants would have products of the unknowns in them. These types of systems are generally very difficult to solve.

So, to avoid this we will do the same thing that we did in the previous example. Everywhere we see a product of constants we will rename it and call it a single constant. The guess that we'll use for this function will be:

$$Y_P(t) = (At^2 + Bt + C)\cos t + (Dt^2 + Et + F)\sin t$$

This is a general rule that we will use when faced with a product of a polynomial and a trig function. We write down the guess for the polynomial and then multiply that by a cosine. We then write down the guess for the polynomial again, using different coefficients, and multiply this by a sine.

$$g(t) = -e^{-2t}(3 - 5t)\cos(9t)$$

This final part has all three parts to it. First, we will ignore the exponential and write down a guess for:

$$-(3 - 5t)\cos(9t)$$

The minus sign can also be ignored. The guess for this is:

$$(At + B)\cos(9t) + (Ct + D)\sin(9t)$$

Now, tack an exponential back on and we're done.

$$Y_P(t) = e^{-2t}(At + B)\cos(9t) + e^{-2t}(Ct + D)\sin(9t)$$

Notice that we put the exponential on both terms.

Example: Find a particular solution for the following differential equation.

$$y'' - 4y' - 12y = 3e^{5t} + \sin(2t) + te^{4t}$$

Solution: This example is the reason that we've been using the same homogeneous differential equation for all the previous examples. There is nothing to do with this problem. All that we need to do it go back to the appropriate examples above and get the particular solution from that example and add them all together.

Doing this gives:

$$Y_P(t) = -\frac{3}{7}e^{5t} + \frac{1}{40}\cos(2t) - \frac{1}{20}\sin(2t) - \frac{1}{36}(3t + 1)e^{4t}$$

Example: Write down the form of the particular solution to,

$$y'' + p(t)y' + q(t)y = g(t)$$

for the following g(t)'s:

$$g(t) = 4\cos(6t) - 9\sin(6t)$$

$$g(t) = -2\sin t + \sin(14t) - 5\cos(14t)$$

$$g(t) = e^{7t} + 6$$

$$g(t) = 6t^2 - 7\sin(3t) + 9$$

$$g(t) = 10e^t - 5te^{-8t} + 2e^{-8t}$$

$$g(t) = t^2 \cos t - 5t \sin t$$

$$g(t) = 5e^{-3t} + e^{-3t}\cos(6t) - \sin(6t)$$

Solution:

$$g(t) = 4\cos(6t) - 9\sin(6t)$$

This first one we've actually already told you how to do. This is in the table of the basic functions. However, we wanted to justify the guess that we put down there. Using the fact on sums of function we would be tempted to write down a guess for the cosine and a guess for the sine. This would give.

$$\underbrace{A\cos(6t) + B\sin(6t)}_{\text{guess for the cosine}} + \underbrace{C\cos(6t) + D\sin(6t)}_{\text{guess for the sine}}$$

So, we would get a cosine from each guess and a sine from each guess. The problem with this as a guess is that we are only going to get two equations to solve after plugging into the differential equation and yet we have 4 unknowns. We will never be able to solve for each of the constants.

To fix this notice that we can combine some terms as follows:

$$(A + C)\cos(6t) + (B + D)\sin(6t)$$

Upon doing this we can see that we've really got a single cosine with a coefficient and a single sine with a coefficient and so we may as well just use:

$$Y_p(t) = A\cos(6t) + B\sin(6t)$$

The general rule of thumb for writing down guesses for functions that involve sums is to always combine like terms into single terms with single coefficients. This will greatly simplify the work required to find the coefficients.

$$g(t) = -2\sin t + \sin(14t) - 5\cos(14t)$$

For this one we will get two sets of sines and cosines. This will arise because we have two different arguments in them. We will get one set for the sine with just a t as its argument and we'll get another set for the sine and cosine with the 14t as their arguments.

The guess for this function is:

$$Y_p(t) = A\cos t + B\sin t + C\cos(14t) + D\sin(14t)$$

$$g(t) = e^{7t} + 6$$

The main point of this problem is dealing with the constant. But that isn't too bad. We just wanted to make sure that an example of that is somewhere in the notes. If you recall that a constant is nothing more than a zero[th] degree polynomial the guess becomes clear.

The guess for this function is:

$$Y_p(t) = Ae^{7t} + B$$

$$g(t) = 6t^2 - 7\sin(3t) + 9$$

This one can be a little tricky if you aren't paying attention. Let's first rewrite the function:

$$g(t) = 6t^2 - 7\sin(3t) + 9$$

$$g(t) = 6t^2 + 9 - 7\sin(3t)$$

All we did was move the 9. However, upon doing that we see that the function is really a sum of a quadratic polynomial and a sine. The guess for this is then:

$$Y_P(t) = At^2 + Bt + C + D\cos(3t) + E\sin(3t)$$

If we don't do this and treat the function as the sum of three terms we would get:

$$At^2 + Bt + C + D\cos(3t) + E\sin(3t) + G$$

and as with the first part in this example we would end up with two terms that are essentially the same (the C and the G) and so would need to be combined. An added step that isn't really necessary if we first rewrite the function.

Look for problems where rearranging the function can simplify the initial guess:

$$g(t) = 10e^t - 5te^{-8t} + 2e^{-8t}$$

So, this look like we've got a sum of three terms here. Let's write down a guess for that:

$$Ae^t + (Bt + C)e^{-8t} + De^{-8t}$$

If we were to multiply the exponential in the second term through we would end up with two terms that are essentially the same and would need to be combined. This is a case where the guess for one term is completely contained in the guess for a different term. When this happens we just drop the guess that's already included in the other term.

So, the guess here is actually:

$$Y_P(t) = Ae^t + (Bt + C)e^{-8t}$$

This arose because we had two terms in our g(t) whose only difference was the polynomial that sat in front of them. When this happens we look at the term that contains the largest degree polynomial, write down the guess for that and don't bother writing down the guess for the other term as that guess will be completely contained in the first guess.

$$g(t) = t^2\cos t - 5t\sin t$$

We've got two terms whose guess without the polynomials in front of them would be the same. Therefore, we will take the one with the largest degree polynomial in front of it and write down the guess for that one and ignore the other term. So, the guess for the function is:

$$Y_P(t) = (At^2 + Bt + C)\cos t + (Dt^2 + Et + F)\sin t$$

$$g(t) = 5e^{-3t} + e^{-3t}\cos(6t) - \sin(6t)$$

This last part is designed to make sure you understand the general rule that we used in the last two parts. This time there really are three terms and we will need a guess for each term. The guess here is:

$$Y_P(t) = Ae^{-3t} + e^{-3t}(B\cos(6t) + C\sin(6t)) + D\cos(6t) + E\sin(6t)$$

We can only combine guesses if they are identical up to the constant. So, we can't combine the first exponential with the second because the second is really multiplied by a cosine and a sine and so the two exponentials are in fact different functions. Likewise, the last sine and cosine can't be combined with those in the middle term because the sine and cosine in the middle term are in fact multiplied by an exponential and so are different.

Example: Find a particular solution for the following differential equation,

$$y'' - 4y' - 12y = e^{6t}$$

Solution: This is especially true given the ease of finding a particular solution for g(t)'s that are just exponential functions. Also, because the point of this example is to illustrate why it is generally a good idea to have the complementary solution in hand first we'll let's go ahead and recall the complementary solution first. Here it is:

$$yc(t) = c_1 e^{-2t} + c_2 e^{6t}$$

Now, without worrying about the complementary solution for a couple more seconds let's go ahead and get to work on the particular solution. There is not much to the guess here. From our previous work we know that the guess for the particular solution should be:

$$Y_P(t) = Ae^{6t}$$

Plugging this into the differential equation gives:

$$36Ae^{6t} - 24Ae^{6t} - 12Ae^{6t} = e^{6t}$$
$$0 = e^{6t}$$

Clearly an exponential can't be zero. So, what went wrong? We finally need the complementary solution. Notice that the second term in the complementary solution is exactly our guess for the form of the particular solution and now recall that both portions of the complementary solution are solutions to the homogeneous differential equation:

$$y'' - 4y' - 12y = 0$$

In other words, we had better have gotten zero by plugging our guess into the differential equation, it is a solution to the homogeneous differential equation. So, how do we fix this? The way that we fix this is to add a tt to our guess as follows:

$$Y_P(t) = Ate^{6t}$$

Plugging this into our differential equation gives:

$$(12Ae^{6t} + 36Ate^{6t}) - 4(Ae^{6t} + 6Ate^{6t}) - 12Ate^{6t} = e^{6t}$$
$$(36A - 24A - 12A)te^{6t} + (12A - 4A)e^{6t} = e^{6t}$$
$$8Ae^{6t} = e^{6t}$$

Now, we can set coefficients equal:

$$8A = 1 \quad \Rightarrow \quad A = \frac{1}{8}$$

So, the particular solution in this case is:

$$Y_P(t) = \frac{t}{8}e^{6t}$$

Example: Write down the guess for the particular solution to the given differential equation. Do not find the coefficients:

$$y'' + 3y' - 28y = 7t + e^{-7t} - 1$$

$$y'' - 100y = 9t^2 e^{10t} + \cos t - t \sin t$$

$$4y'' + y = e^{-2t} \sin\left(\frac{t}{2}\right) + 6t \cos\left(\frac{t}{2}\right)$$

$$4y'' + 16y' + 17y = e^{-2t} \sin\left(\frac{t}{2}\right) + 6t \cos\left(\frac{t}{2}\right)$$

$$y'' + 8y' + 16y = e^{-4t} + (t^2 + 5)e^{-4t}$$

Solution: In these solutions we'll leave the details of checking the complementary solution to you,

$$y'' + 3y' - 28y = 7t + e^{-7t} - 1$$

The complementary solution is:

$$y_c(t) = c_1 e^{4t} + c_2 e^{-7t}$$

Remembering to put the "-1" with the 7t gives a first guess for the particular solution:

$$Y_P(t) = At + B + Ce^{-7t}$$

The last term in the guess is the last term in the complementary solution. The first two terms however aren't a problem and don't appear in the complementary solution. Therefore, we will only add a tt onto the last term.

The correct guess for the form of the particular solution is:

$$Y_P(t) = At + B + Cte^{-7t}$$

$$y'' - 100y = 9t^2 e^{10t} + \cos t - t \sin t$$

The complementary solution is:

$$y_c(t) = c_1 e^{10t} + c_2 e^{-10t}$$

A first guess for the particular solution is:

$$Y_P(t) = (At^2 + Bt + C)e^{10t} + (Et + F)\cos t + (Gt + H)\sin t$$

If we multiplied the exponential term through the parenthesis that we would end up getting part of the complementary solution showing up. Since the problem part arises from the first term the *whole* first term will get multiplied by tt. The second and third terms are okay as they are.

The correct guess for the form of the particular solution in this case is:

$$Y_P(t) = t(At^2 + Bt + C)e^{10t} + (Et + F)\cos t + (Gt + H)\sin t$$

So, in general, if you were to multiply out a guess and if any term in the result shows up in the complementary solution, then the whole term will get a tt not just the problem portion of the term.

$$4y'' + y = e^{-2t}\sin\left(\frac{t}{2}\right) + 6t\cos\left(\frac{t}{2}\right)$$

The complementary solution is:

$$y_c(t) = c_1 \cos\left(\frac{t}{2}\right) + c_2 \sin\left(\frac{t}{2}\right)$$

A first guess for the particular solution is:

$$Y_P(t) = e^{-2t}\left(A\cos\left(\frac{t}{2}\right) + B\sin\left(\frac{t}{2}\right)\right) + (Ct + D)\cos\left(\frac{t}{2}\right) + (Et + F)\sin\left(\frac{t}{2}\right)$$

In this case both the second and third terms contain portions of the complementary solution. The first term doesn't however, since upon multiplying out, both the sine and the cosine would have an exponential with them and that isn't part of the complementary solution. We only need to worry about terms showing up in the complementary solution if the only difference between the complementary solution term and the particular guess term is the constant in front of them.

So, in this case the second and third terms will get a tt while the first won't.

The correct guess for the form of the particular solution is:

$$Y_P(t) = e^{-2t}\left(A\cos\left(\frac{t}{2}\right) + B\sin\left(\frac{t}{2}\right)\right) + t(Ct + D)\cos\left(\frac{t}{2}\right) + t(Et + F)\sin\left(\frac{t}{2}\right)$$

$$4y'' + 16y' + 17y = e^{-2t}\sin\left(\frac{t}{2}\right) + 6t\cos\left(\frac{t}{2}\right)$$

To get this problem we changed the differential equation from the last example and left the g(t) alone. The complementary solution this time is:

$$y_c(t) = c_1 e^{-2t} \cos\left(\frac{t}{2}\right) + c_2 e^{-2t} \sin\left(\frac{t}{2}\right)$$

As with the last part, a first guess for the particular solution is:

$$Y_P(t) = e^{-2t}\left(A\cos\left(\frac{t}{2}\right) + B\sin\left(\frac{t}{2}\right)\right) + (Ct + D)\cos\left(\frac{t}{2}\right) + (Et + F)\sin\left(\frac{t}{2}\right)$$

This time however it is the first term that causes problems and not the second or third. In fact, the first term is exactly the complementary solution and so it will need a t. Recall that we will only have a problem with a term in our guess if it only differs from the complementary solution by a constant. The second and third terms in our guess don't have the exponential in them and so they don't differ from the complementary solution by only a constant.

The correct guess for the form of the particular solution is:

$$Y_P(t) = te^{-2t}\left(A\cos\left(\frac{t}{2}\right) + B\sin\left(\frac{t}{2}\right)\right) + (Ct + D)\cos\left(\frac{t}{2}\right) + (Et + F)\sin\left(\frac{t}{2}\right)$$

$$y'' + 8y' + 16y = e^{-4t} + (t^2 + 5)e^{-4t}$$

The complementary solution is:

$$y_c(t) = c_1 e^{-4t} + c_2 te^{-4\tau}$$

The two terms in g(t) are identical with the exception of a polynomial in front of them. So this means that we only need to look at the term with the highest degree polynomial in front of it. A first guess for the particular solution is:

$$Y_P(t) = (At^2 + Bt + C)e^{-4t}$$

Notice that if we multiplied the exponential term through the parenthesis the last two terms would be the complementary solution. Therefore, we will need to multiply this whole thing by a t.

The next guess for the particular solution is then:

$$Y_P(t) = t(At^2 + Bt + C)e^{-4t}$$

This still causes problems however. If we multiplied the tt and the exponential through, the last term will still be in the complementary solution. In this case, unlike the previous ones, a t wasn't sufficient to fix the problem. So, we will add in another t to our guess.

The correct guess for the form of the particular solution is:

$$Y(t) = t^2(At^2 + Bt + C)e^{-4}$$

Upon multiplying this out none of the terms are in the complementary solution and so it will be okay.

Variation of Parameters

Sometimes, r(x) is not a combination of polynomials, exponentials, or sines and cosines. When this is the case, the method of undetermined coefficients does not work, and we have to use another approach to find a particular solution to the differential equation. We use an approach called the method of variation of parameters.

To simplify our calculations a little, we are going to divide the differential equation through by a, so we have a leading coefficient of 1. Then the differential equation has the form:

$$y'' + py' + qy = r(x),$$

where p and q are constants.

If the general solution to the complementary equation is given by $c_1 y_1(x) + c_2 y_2(x)$, we are going to look for a particular solution of the form:

$$y_p(x) = u(x) y_1(x) + v(x) y_2(x).$$

In this case, we use the two linearly independent solutions to the complementary equation to form our particular solution. However, we are assuming the coefficients are functions of xx, rather than constants. We want to find functions $u(x)$ and $v(x)$ such that $y_p(x)$ satisfies the differential equation. We have:

$$y_p = uy_1 + vy_2$$
$$y_p' = u'y_1 + uy_1' + v'y_2 + vy_2'$$
$$y_p'' = (u'y_1 + v'y_2)' + u'y_1' + uy_1'' + v'y_2' + vy_2''.$$

Substituting into the differential equation, we obtain:

$$y_p'' + py_p' + qy_p = [(u'y_1 + v'y_2)' + u'y_1' + uy_1'' + v'y_2' + vy_2'']$$
$$+ p[u'y_1 + uy_1' + v'y_2 + vy_2'] + q[uy_1 + vy_2]$$
$$= u[y_1'' + p_y 1' + qy_1] + v[y_2'' + py_2' + qy_2]$$
$$+ (u'y_1 + v'y_2)' + p(u'y_1 + v'y_2) + (u'y_1' + v'y_2').$$

Note that y_1 and y_2 are solutions to the complementary equation, so the first two terms are zero. Thus, we have:

$$(u'y_1 + v'y_2)' + p(u'y_1 + v'y_2) + (u'y_1' + v'y_2') = r(x).$$

If we simplify this equation by imposing the additional condition $u'y_1 + v'y_2 = 0$, the first two terms are zero, and this reduces to $u'y_1' + v'y_2' = r(x)$. So, with this additional condition, we have a system of two equations in two unknowns:

$$u'y_1 + v'y_2 = 0$$
$$u'y_1' + v'y_2' = r(x).$$

Solving this system gives us u' and v', which we can integrate to find u and v.

Then, $y_p(x) = u(x)y_1(x) + v(x)y_2(x)$ is a particular solution to the differential equation. Solving this system of equations is sometimes challenging and requires the knowledge of Cramer's rule.

Cramer's Rule

The system of equations:

$$a_1 z_1 + b_1 z_2 = r_1$$
$$a_2 z_1 + b_2 z_2 = r_2$$

has a unique solution if and only if the determinant of the coefficients is not zero. In this case, the solution is given by:

$$z_1 = \frac{\begin{vmatrix} r_1 & b_1 \\ r_2 & b_2 \end{vmatrix}}{\begin{vmatrix} a_1 & b_1 \\ a_2 & b_2 \end{vmatrix}} \quad \text{and} \quad z_2 = \frac{\begin{vmatrix} a_1 & r_1 \\ a_2 & r_2 \end{vmatrix}}{\begin{vmatrix} a_1 & b_1 \\ a_2 & b_2 \end{vmatrix}}.$$

Example: Using Cramer's Rule.

Use Cramer's rule to solve the following system of equations.

$$x^2 z_1 + 2x z_2 = 0$$
$$z_1 - 3x^2 z_2 = 2x$$

Solution:

We have,

$$a_1(x) = x^2$$
$$a_2(x) = 1$$
$$b_1(x) = 2x$$
$$b_2(x) = -3x^2$$
$$r_1(x) = 0$$
$$r_2(x) = 2x.$$

Then,

$$\begin{vmatrix} a_1 & b_1 \\ a_2 & b_2 \end{vmatrix} = \begin{vmatrix} x^2 & 2x \\ 1 & -3x^2 \end{vmatrix} = -3x^4 - 2x$$

and

$$\begin{vmatrix} r_1 & b_1 \\ r_2 & b_2 \end{vmatrix} = \begin{vmatrix} 0 & 2x \\ 2x & -3x^2 \end{vmatrix} = 0 - 4x^2 = -4x^2.$$

Thus,

$$z1 = \frac{\begin{vmatrix} r_1 & b_1 \\ r_2 & b_2 \end{vmatrix}}{\begin{vmatrix} a_1 & b_1 \\ a_2 & b_2 \end{vmatrix}} = \frac{-4x^2}{-3x^4 - 2x} = \frac{4x}{3x^3 + 2}.$$

In addition,

$$\begin{vmatrix} a_1 & r_1 \\ a_2 & r_2 \end{vmatrix} = \begin{vmatrix} x^2 & 0 \\ 1 & 2x \end{vmatrix} = 2x^3 - 0 = 2x^3.$$

Thus,

$$z2 = \frac{\begin{vmatrix} a_1 & r_1 \\ a_2 & r_2 \end{vmatrix}}{\begin{vmatrix} a_1 & b_1 \\ a_2 & b_2 \end{vmatrix}} = \frac{2x^3}{-3x^4 - 2x} = \frac{-2x^2}{3x^3 + 2}.$$

Problem-Solving Strategy: Method of Variation of Parameters

- Solve the complementary equation and write down the general solution

$$c_1 y_1(x) + c_2 y_2(x).$$

- Use Cramer's rule or another suitable technique to find functions u'(x) and v'(x) satisfying

$$u'y_1 + v'y_2 = 0$$
$$u'y_1' + v'y_2' = r(x).$$

- Integrate u' and v' to find u(x) and v(x). Then, $y_p(x) = u(x)y_1(x) + v(x)y_2(x)$ is a particular solution to the equation.

- Add the general solution to the complementary equation and the particular solution found in step 3 to obtain the general solution to the nonhomogeneous equation.

Example: Using the Method of Variation of Parameters.

Find the general solution to the following differential equations:

$$y'' - 2y' + y = \frac{e^t}{t^2}$$

$$y'' + y = 3\sin^2 x$$

Solution:

The complementary equation is $y'' - 2y' + y = 0$ with associated general solution $c_1 e^t + c_2 t e^t$.

Therefore, $y_1(t) = e^t$ and $y_2(t) = te^t$. Calculating the derivatives, we get $y_1'(t) = e^t$

and $y_2{}'(t) = e^t + te^t$. Then, we want to find functions u'(t) and v'(t) so that:

$$u'e^t + v'te^t = 0$$

$$u'e^t + v'(e^t + te^t) = \frac{e^t}{t^2}.$$

Applying Cramer's rule (equation $z_1 = \dfrac{\begin{vmatrix} r_1 & b_1 \\ r_2 & b_2 \end{vmatrix}}{\begin{vmatrix} a_1 & b_1 \\ a_2 & b_2 \end{vmatrix}}$ and $z_2 = \dfrac{\begin{vmatrix} a_1 & r_1 \\ a_2 & r_2 \end{vmatrix}}{\begin{vmatrix} a_1 & b_1 \\ a_2 & b_2 \end{vmatrix}}$), we have:

$$u' = \frac{\begin{vmatrix} 0 & te^t \\ \dfrac{e^t}{t^2} & e^t + te^t \end{vmatrix}}{\begin{vmatrix} e^t & te^t \\ e^t & e^t + te^t \end{vmatrix}} = \frac{0 - te^t(\dfrac{e^t}{t^2})}{e^t(e^t + te^t) - e^t te^t} = \frac{-\dfrac{e^{2t}}{t}}{e^{2t}} = -\frac{1}{t}$$

and

$$v' = \frac{\begin{vmatrix} e^t & 0 \\ e^t & \dfrac{e^t}{t^2} \end{vmatrix}}{\begin{vmatrix} e^t & te^t \\ e^t & e^t + te^t \end{vmatrix}} = \frac{e^t(\dfrac{e^t}{t^2})}{e^{2t}} = \frac{1}{t^2}.$$

Integrating, we get:

$$u = -\int \frac{1}{t} dt = -\ln|t|$$

$$v = \int \frac{1}{t^2} dt = -\frac{1}{t}$$

Then we have,

$$y_p = -e^t \ln|t| - \frac{1}{t} te^t$$

$$= -e^t \ln|t| - e^t.$$

The e^t term is a solution to the complementary equation, so we don't need to carry that term into our general solution explicitly. The general solution is:

$$y(t) = c_1 e^t + c_2 te^t - e^t \ln|t|$$

The complementary equation is $y''+y=0$ with associated general solution $c_1 \cos x + c_2 \sin x$. So, $y_1(x) = \cos x$ and $y_2(x) = \sin x$. Then, we want to find functions u'(x) and v'(x) such that:

$$u'\cos x + v'\sin x = 0$$
$$-u'\sin x + v'\cos x = 3\sin_2 x.$$

Applying Cramer's rule, we have:

$$u' = \frac{\begin{vmatrix} 0 & \sin x \\ 3\sin^2 x & \cos x \end{vmatrix}}{\begin{vmatrix} \cos x & \sin x \\ -\sin x & \cos x \end{vmatrix}} = \frac{0 - 3\sin^3 x}{\cos^2 x + \sin^2 x} = -3\sin^3 x$$

and

$$v' = \frac{\begin{vmatrix} \cos x & 0 \\ -\sin x & 3\sin^2 x \end{vmatrix}}{\begin{vmatrix} \cos x & \sin x \\ -\sin x & \cos x \end{vmatrix}} = \frac{3\sin^2 x \cos x}{1} = 3\sin^2 x \cos x.$$

Integrating first to find u, we get:

$$u = \int -3\sin^3 x dx = -3\left[-\frac{1}{3}\sin^2 x \cos x + \frac{2}{3}\int \sin x dx \right] = \sin^2 x \cos x + 2\cos x.$$

Now, we integrate to find v. Using substitution (with $w = \sin x$), we get:

$$v = \int 3\sin^2 x \cos x dx = \int 3w^2 dw = w^3 = \sin^3 x.$$

Then,

$$y_p = (\sin^2 x \cos x + 2\cos x)\cos x + (\sin^3 x)\sin x$$
$$= \sin_2 x \cos_2 x + 2\cos_2 x + \sin_4 x$$
$$= 2\cos_2 x + \sin_2 x(\cos^2 x + \sin^2 x)$$
$$= 2\cos_2 x + \sin_2 x$$
$$= \cos_2 x + 1$$

The general solution is:

$$y(x) = c_1 \cos x + c_2 \sin x + 1 + \cos^2 x.$$

Example: Find a general solution to the following differential equation,

$$2y'' + 18y = 6\tan(3t)$$

Solution: First, since the formula for variation of parameters requires a coefficient of a one in front of the second derivative let's take care of that before we forget. The differential equation that we'll actually be solving is:

$$y'' + 9y = 3\tan(3t)$$

We'll leave it to you to verify that the complementary solution for this differential equation is:

$$y_c(t) = c_1 \cos(3t) + c_2 \sin(3t)$$

So, we have:

$$y_1(t) = \cos(3t) \qquad y_2(t) = \sin(3t)$$

The Wronskian of these two functions is:

$$W = \begin{vmatrix} \cos(3t) & \sin(3t) \\ -3\sin(3t) & 3\cos(3t) \end{vmatrix} = 3cos^2(3t) + 3sin^2(3t) = 3$$

The particular solution is then,

$$Y_p(t) = -\cos(3t)\int \frac{3\sin(3t)\tan(3t)}{3} dt + \sin(3t)\int \frac{3\cos(3t)\tan(3t)}{3} dt$$

$$= -\cos(3t)\int \frac{\sin^2(3t)}{\cos(3t)} dt + \sin(3t)\int \sin(3t)dt$$

$$= -\cos(3t)\int \frac{1-\cos^2(3t)}{\cos(3t)} dt + \sin(3t)\int \sin(3t)dt$$

$$= -\cos(3t)\int \sec(3t) - \cos(3t)dt + \sin(3t)\int \sin(3t)dt$$

$$= -\frac{\cos(3t)}{3}(\ln|\sec(3t)+\tan(3t)|-\sin(3t)) + \frac{\sin(3t)}{3}(-\cos(3t))$$

$$= -\frac{\cos(3t)}{3}\ln|\sec(3t)+\tan(3t)|$$

The general solution is:

$$y(t) = c_1 \cos(3t) + c_2 \sin(3t) - \frac{\cos(3t)}{3}\ln|\sec(3t)+\tan(3t)|$$

Example: Find a general solution to the following differential equation,

$$y'' - 2y' + y = \frac{e^t}{t^2+1}$$

Solution: We first need the complementary solution for this differential equation. We'll leave it to you to verify that the complementary solution is:

$$y_c(t) = c_1 e^t + c_2 t e^t$$

So, we have:

$$y_1(t) = e^t \qquad y_2(t) = te^t$$

The Wronskian of these two functions is:

$$W = \begin{vmatrix} e^t & te^t \\ e^t & e^t + te^t \end{vmatrix} = e^t(e^t + te^t) - e^t(te^t) = e^{2\tau}$$

The particular solution is then,

$$Y_P(t) = -e^t \int \frac{te^t\ e^t}{e^{2t}(t^2+1)} dt + te^t \int \frac{e^t\ e^t}{e^{2t}(t^2+1)} dt$$

$$= -e^t \int \frac{t}{t^2+1} dt + te^t \int \frac{1}{t^2+1} dt$$

$$= -\frac{1}{2} e^t \ln(1+t^2) + te^t \tan^{-1}(t)$$

The general solution is:

$$y(t) = c_1 e^t + c_2 te^t - \frac{1}{2} e^t \ln(1+t^2) + te^t \tan^{-1}(t)$$

Example: Find the general solution to,

$$ty'' - (t+1)y' + y = t^2$$

given that:

$$y_1(t) = e^t \qquad y_2(t) = t+1$$

form a fundamental set of solutions for the homogeneous differential equation.

Solution: As with the first example, we first need to divide out by a t.

$$y'' - \left(1 + \frac{1}{t}\right) y' + \frac{1}{t} y = t$$

The Wronskian for the fundamental set of solutions is:

$$W = \begin{vmatrix} e^t & t+1 \\ e^t & 1 \end{vmatrix} = e^t - e^t(t+1) = -te^t$$

The particular solution is:

$$Y_P(t) = -e^t \int \frac{(t+1)t}{-te^t} dt + (t+1) \int \frac{e^t(t)}{-te^t} dt$$

$$= e^t \int (t+1)e^{-t} dt - (t+1) \int dt$$
$$= e^t (-e^{-t}(t+2)) - (t+1)t$$
$$= -t^2 - 2t - 2$$

The general solution for this differential equation is:

$$y(t) = c_1 e + c_2(t+1) - t - 2t - 2$$

Higher Order Linear Homogeneous Differential Equations with Constant Coefficients

The linear homogeneous differential equation of the nth order with constant coefficients can be written as:

$$y^{(n)}(x) + a_1 y^{(n-1)}(x) + \cdots + a_{n-1} y'(x) + a_n y(x) = 0,$$

where $a_1, a_2, \ldots, a_n$ are constants which may be real or complex.

Using the linear differential operator $L(D)$, this equation can be represented as:

$$L(D)y(x) = 0,$$

where,

$$L(D) = D^n + a_1 D^{n-1} + \cdots + a_{n-1} D + a_n.$$

For each differential operator with constant coefficients, we can introduce the characteristic polynomial:

$$L(\lambda) = \lambda^n + a_1 \lambda^{n-1} + \cdots + a_{n-1} \lambda + a_n.$$

The algebraic equation:

$$L(\lambda) = \lambda^n + a_1 \lambda^{n-1} + \cdots + a_{n-1} \lambda + a_n = 0$$

is called the characteristic equation of the differential equation.

According to the fundamental theorem of algebra, a polynomial of degree n has exactly n roots, counting multiplicity. In this case the roots can be both real and complex (even if all the coefficients of $a_1, a_2, \ldots, a_n$ are real).

Let us consider in more detail the different cases of the roots of the characteristic equation and the corresponding formulas for the general solution of differential equations.

Case: All roots of the characteristic equation are real and distinct.

We assume that the characteristic equation $L(\lambda)=0$ has n roots $\lambda_1, \lambda_2, \ldots, \lambda_n$. In this case the general solution of the differential equation is written in a simple form:

$$y(x) = C_1 e^{\lambda_1 x} + C_2 e^{\lambda_2 x} + \cdots + C_n e^{\lambda_n x},$$

where C_1, C_2,..., C_n are constants depending on initial conditions.

Case: The roots of the characteristic equation are real and multiple.

Let the characteristic equation $L(\lambda)=0$ of degree n have m roots $\lambda_1, \lambda_2, \ldots, \lambda_m$, the multiplicity of which, respectively, is equal to $k_1, k_2, \ldots, k_m$. It is clear that the following condition holds:

$$k_1 + k_2 + \cdots + k_m = n.$$

Then the general solution of the homogeneous differential equations with constant coefficients has the form:

$$y(x) = C_1 e^{\lambda_1 x} + C_2 x e^{\lambda_1 x} + \cdots + C_{k1} x^{k_1-1} e^{\lambda_1 x} + \cdots + C_{n-k_m+1} e^{\lambda_m x}$$
$$+ C_{n-k_m+2} x e^{\lambda_m x} + \cdots + C_n x^{k_m-1} e^{\lambda_m x}.$$

It is seen that the formula of the general solution has exactly k_i terms corresponding to each root λ_i of multiplicity k_i. These terms are formed by multiplying x to a certain degree by the exponential function $e^{\lambda_i x}$. The degree of x varies in the range from 0 to $k_i - 1$, where k_i is the multiplicity of the root λ_i.

Case: The roots of the characteristic equation are complex and distinct.

If the coefficients of the differential equation are real numbers, the complex roots of the characteristic equation will be presented in the form of conjugate pairs of complex numbers:

$$\lambda_{1,2} = \alpha \pm i\beta, \lambda_{3,4} = \gamma \pm i\delta, \ldots$$

In this case the general solution is written as:

$$y(x) = e^{\alpha x}(C_1 \cos \beta x + C_2 \sin \beta x) + e \quad (C_3 \cos \delta x + C_4 \sin \delta x) + \cdots$$

Case: The roots of the characteristic equation are complex and multiple.

Here, each pair of complex conjugate roots $\alpha \pm i\beta$ of multiplicity k produces 2k particular solutions:

$$e^{\alpha x} \cos \beta x, e^{\alpha x} \sin \beta x, e^{\alpha x} x \cos \beta x, e^{\alpha x} x \sin \beta x, \ldots, e^{\alpha x} x^{k-1} \cos \beta x,$$
$$e^{\alpha x} x^{k-1} \sin \beta x.$$

Then the part of the general solution of the differential equation corresponding to a given pair of complex conjugate roots is constructed as follows:

$$y(x) = e^{\alpha x}(C_1 \cos \beta x + C_2 \sin \beta x) + x e^{\alpha x}(C_3 \cos \beta x + C_4 \sin \beta x) + \cdots$$
$$+ x^{k-1} e^{\alpha x}(C_{2k-1} \cos \beta x + C_{2k} \sin \beta x).$$

In general, when the characteristic equation has both real and complex roots of arbitrary multiplicity, the general solution is constructed as the sum of the above solutions of the form 1–4.

Example: Solve the differential equation $y''' + 2y'' - y' - 2y = 0$.

Solution: Write the corresponding characteristic equation:

$$\lambda^3 + 2\lambda^2 - \lambda - 2 = 0.$$

Solving it, we find the roots:

$$\lambda^2(\lambda + 2) - (\lambda + 2) = 0, \Rightarrow (\lambda + 2)(\lambda^2 - 1) = 0,$$
$$\Rightarrow (\lambda + 2)(\lambda - 1)(\lambda + 1) = 0, \Rightarrow \lambda_1 = -2, \lambda_2 = 1, \lambda_3 = -1.$$

It is seen that all three roots are real. Therefore, the general solution of the differential equations can be written as:

$$y(x) = C_1 e^{-2x} + C_2 e^x + C_3 e^{-x},$$

where C_1, C_2, C_3 are arbitrary constants.

Example: Solve the equation $y''' - 7y'' + 11y' - 5y = 0$.

Solution: The corresponding characteristic equation is:

$$\lambda^3 - 7\lambda^2 + 11\lambda - 5 = 0.$$

It is easy to see that one of the roots is the number λ=1. Then, factoring the term (λ–1) from the equation, we obtain:

$$\lambda^3 - \lambda^2 - 6\lambda^2 + 6\lambda + 5\lambda - 5 = 0, \Rightarrow \lambda^2(\lambda - 1) - 6\lambda(\lambda - 1) + 5(\lambda - 1) = 0,$$
$$\Rightarrow (\lambda - 1) \cdot (\lambda^2 - 6\lambda + 5) = 0, \Rightarrow (\lambda - 1) \cdot (\lambda - 1) \cdot (\lambda - 5) = 0,$$
$$\Rightarrow (\lambda - 1)^2(\lambda - 5) = 0.$$

Thus, the equation has two roots $\lambda_1 = 1, \lambda_2 = 5,$ the first of which has multiplicity 2. Then the general solution of differential equations can be written as follows:

$$y(x) = (C_1 + C_2 x)e^x + C_3 e^{5x},$$

where C_1, C_2, C_3 are arbitrary numbers.

Example: Solve the equation $y^{IV} - y''' + 2y' = 0$.

Solution: Write the characteristic equation:

$$\lambda^4 - \lambda^3 + 2\lambda = 0.$$

Factor the left side and find the roots:

$$\lambda(\lambda^3 - \lambda^2 + 2) = 0.$$

One of the roots of the cubic polynomial is the number $\lambda = -1$. Therefore, we divide $\lambda^3 - \lambda^2 + 2$ by $\lambda + 1$:

$$\frac{\lambda^3 - \lambda^2 + 2}{\lambda + 1} = \lambda^2 - 2\lambda + 2.$$

As a result, the characteristic equation takes the following form:

$$\lambda(\lambda + 1) \cdot (\lambda^2 - 2\lambda + 2) = 0.$$

We find the roots of the quadratic equation:

$$\lambda^2 - 2\lambda + 2 = 0, \Rightarrow D = 4 - 8 = -4, \Rightarrow \lambda = \frac{2 \pm \sqrt{-4}}{2}$$

$$= \frac{2 \pm 2i}{2} = 1 \pm i.$$

Thus, the characteristic equation has four distinct roots, two of which are complex:

$$\lambda_1 = 0, \lambda_2 = -1, \lambda_{3,4} = 1 \pm i.$$

The general solution of the differential equation can be represented as:

$$y(x) = C_1 + C_2 e^{-x} + e^x (C_3 \cos x + C_4 \sin x),$$

where $C_1, \ldots, C_4$ are arbitrary constants.

Example: Solve the equation $y^V + 18y''' + 81y' = 0$.

Solution: The characteristic equation can be written as:

$$\lambda^5 + 18\lambda^3 + 81\lambda = 0.$$

Factor the left side and calculate the roots:

$$\lambda(\lambda^4 + 18\lambda^2 + 81) = 0, \Rightarrow \lambda(\lambda^2 + 9)^2 = 0.$$

As it can be seen, the equation has the following roots:

$$\lambda_1 = 0, \lambda_{2,3} = \pm 3i,$$

and imaginary roots have multiplicity 2. In accordance with the rules set out above, we write the general solution in the form:

$$y(x) = C_1 + (C_2 + C_3 x)\cos 3x + (C_4 + C_5 x)\sin 3x,$$

where $C_1, \ldots, C_5$ are arbitrary numbers.

Example: Solve the differential equation $y^{IV} - 4y''' + 5y'' - 4y' + 4y = 0$.

Solution: Calculate the roots of the characteristic equation,

$$\lambda^4 - 4\lambda^3 + 5\lambda^2 - 4\lambda + 4 = 0.$$

Factor the left side:

$$\lambda^4 - 2\lambda^3 - 2\lambda^3 + 4\lambda^2 + \lambda^2 - 2\lambda - 2\lambda + 4 = 0,$$
$$\Rightarrow (\lambda^4 - 2\lambda^3) - (2\lambda^3 - 4\lambda^2) + (\lambda^2 - 2\lambda) - (2\lambda - 4) = 0,$$
$$\Rightarrow \lambda^3(\lambda - 2) - 2\lambda^2(\lambda - 2) + \lambda(\lambda - 2) - 2(\lambda - 2) = 0,$$
$$\Rightarrow (\lambda - 2) \cdot (\lambda^3 - 2\lambda^2 + \lambda - 2) = 0,$$
$$\Rightarrow (\lambda - 2) \cdot [\lambda^2(\lambda - 2) + \lambda - 2] = 0,$$
$$\Rightarrow (\lambda - 2) \cdot (\lambda - 2) \cdot (\lambda^2 + 1) = 0, \Rightarrow (\lambda - 2)^2(\lambda^2 + 1) = 0.$$

We see that the roots of the equation are equal:

$$\lambda_1 = 2, \lambda_{3,4} = \pm i.$$

The first root is of multiplicity 2. The general solution of the differential equation is given by:

$$y(x) = (C_1 + C_2 x)e^{2x} + C_3 \cos x + C_4 \sin x,$$

where $C_1, \ldots, C_4$ are as usual arbitrary constants.

Higher Order Linear Homogeneous Differential Equations with Variable Coefficients

The linear homogeneous equation of the nth order has the form:

$$y^{(n)} + a_1(x)y^{(n-1)} + \cdots + a_{n-1}(x)y' + a_n(x)y = 0,$$

where the coefficients $a_1(x), a_2(x), \ldots, a_n(x)$ are continuous functions on some interval $[a,b]$.

The left side of the equation can be written in abbreviated form using the linear differential operator L:

$$Ly(x) = 0,$$

where L denotes the set of operations of differentiation, multiplication by the coefficients $a_i(x)$, and addition.

The operator L is linear, and therefore has the following properties:

$$L[y_1(x) + y_2(x)] = L[y_1(x)] + L[y_2(x)],$$

$$L[Cy(x)] = CL[y(x)],$$

where $y_1(x) y_2(x)$ are arbitrary, $n-1$ times differentiable functions, C is any number.

It follows from the properties of the operator L that if the functions $y_1, y_2, \ldots, y_n$ are solutions of the homogeneous differential equation of the nth order, then the function of the form:

$$y(x) = C_1 y_1 + C_2 y_2 + \cdots + C_n y_n,$$

where $C_1, C_2, \ldots, C_n$ are arbitrary constants, will also satisfy this equation.

The last expression is the general solution of homogeneous differential equation if the functions $y_1, y_2, \ldots, y_n$ form a fundamental system of solutions.

Fundamental System of Solutions

The set of n linearly independent particular solutions $y_1, y_2, \ldots, y_n$ is called a fundamental system of the homogeneous linear differential equation of the nth order.

The functions $y_1, y_2, \ldots, y_n$ are linearly independent on the interval $[a, b]$. If the identity:

$$\alpha_1 y_1 + \alpha_2 y_2 + \cdots + \alpha_n y_n \equiv 0$$

holds only provided:

$$\alpha_1 = \alpha_2 = \cdots = \alpha_n = 0,$$

where the numbers $\alpha_1, \alpha_2, \ldots, \alpha_n$ are not simultaneously 0.

To test functions for linear independence it is convenient to use the Wronskian:

$$W(x) = W_{y_1, y_2, \ldots, y_n}(x) = \begin{vmatrix} y_1 & y_2 & \cdots & y_n \\ y_1' & y_2' & \cdots & y_n' \\ \cdots & \cdots & \cdots & \cdots \\ y_1^{(n-1)} & y_2^{(n-1)} & \cdots & y_n^{(n-1)} \end{vmatrix}.$$

Let the functions $y_1, y_2, \ldots, y_n$ be $n-1$ times differentiable on the interval $[a, b]$. Then if these functions are linearly dependent on the interval $[a, b]$, then the following identity holds:

$$W(x) \equiv 0.$$

Accordingly, if these functions are linearly independent on $[a, b]$, we have the formula:

$$W(x) \neq 0.$$

The fundamental system of solutions uniquely defines a linear homogeneous differential equation. In particular, the fundamental system y_1, y_2, y_3 defines a third-order equation, which is expressed through determinant as follows:

$$\begin{vmatrix} y_1 & y_2 & y_3 & y \\ y_1' & y_2' & y_3' & y' \\ y_1'' & y_2'' & y_3'' & y'' \\ y_1''' & y_2''' & y_3''' & y''' \end{vmatrix} = 0.$$

The expression for the differential equation of the nth order can be written similarly:

$$\begin{vmatrix} y_1 & y_2 & \cdots & y_n & y \\ y_1' & y_2' & \cdots & y_n' & y' \\ \cdots & \cdots & \cdots & \cdots & \cdots \\ y_1^{(n)} & y_2^{(n)} & \cdots & y_n^{(n)} & y^{(n)} \end{vmatrix} = 0.$$

Liouville's Formula

Suppose that the functions $y_1, y_2, \ldots, y_n$ form a fundamental system of solutions for a differential equations of nth order. Suppose that the point x_0 belongs to the interval $[a, b]$. Then the Wronskian is determined by Liouville's formula:

$$W(x) = W(x_0) e^{-\int_{x_0}^{x} a_1(t)dt},$$

where α_1 is the coefficient of the derivative $y^{(n-1)}$ in the differential equation. Here we assume that the coefficient $a_0(x)$ of $y^{(n)}$ in the differential equation is equal to 1. Otherwise, Liouville's formula takes the form:

$$W(x) = W(x_0) e^{-x\int_{x_0}^{x} \frac{a_1(t)}{a_0(t)}dt}, a_0(t) \neq 0, t \in [a, b].$$

Reduction of Order of a Homogeneous Linear Equation

The order of a linear homogeneous equation:

$$Ly(x) = y^{(n)} + a_1(x)y^{(n-1)} + \cdots + a_{n-1}(x)y' + a_n(x)y = 0$$

can be reduced by one by the substitution $y' = yz$. Unfortunately, usually such a substitution does not simplify the solution, because the new equation in the variable z becomes nonlinear.

If a particular solution y_1 is known, then the order of the differential equation can be reduced (while maintaining its linearity) by replacing:

$$y = y_1 z, \; z' = u.$$

In general, if we know k linearly independent particular solutions, the order of the equation can be reduced by k units.

Example: Show that the functions $x, \sin x, \cos x$ are linearly independent.

Solution: We find the Wronskian matrix $W(x)$ for this system of functions:

$$W(x) = \begin{vmatrix} x & \sin x & \cos x \\ 1 & \cos x & -\sin x \\ 0 & -\sin x & -\cos x \end{vmatrix} = x \begin{vmatrix} \cos x & -\sin x \\ -\sin x & -\cos x \end{vmatrix} - 1. \begin{vmatrix} \sin x & \cos x \\ -\sin x & -\cos x \end{vmatrix}$$

$$= x\left(-\cos^2 x - \sin^2 x\right) - 1.\left(-\sin x \cos x + \sin x \cos x\right) = -x \neq 0.$$

Since the Wronskian is not identically zero, it follows that the given system of functions is linearly independent.

Example: Show that the functions x, x^2, x^3, x^4 form a linearly independent system.

Solution: We compute the corresponding Wronskian,

$$W(x) = \begin{vmatrix} x & x^2 & x^3 & x^4 \\ 1 & 2x & 3x^2 & 4x^3 \\ 0 & 2 & 6x & 12x^2 \\ 0 & 0 & 6 & 24x \end{vmatrix} R_1 - xR_2 = \left(-\frac{1}{x}\right) \cdot \begin{vmatrix} x & x^2 & x^3 & x^4 \\ 0 & -x^2 & -2x^3 & -3x^4 \\ 0 & 2 & 6x & 12x^2 \\ 0 & 0 & 6 & 24x \end{vmatrix}$$

$$= \left(-\frac{1}{x}\right) \cdot x \cdot (-1) \cdot \begin{vmatrix} x^2 & 2x^3 & 3x^4 \\ 2 & 6x & 12x^2 \\ 0 & 6 & 24x \end{vmatrix} = x^2 \begin{vmatrix} 6x & 12x^2 \\ 6 & 24x \end{vmatrix} - 2 \begin{vmatrix} 2x^3 & 3x^4 \\ 6 & 24x \end{vmatrix}$$

$$= x^2(144x^2 - 72x^2) - 2(48x^4 - 18x^4) = 12x^4 \neq 0.$$

As the determinant is not identically equal to zero, these functions are linearly independent.

Example: Make a differential equation, which is determined by the fundamental system of functions $1, x^2, e^x$.

Solution: This equation is written in terms of the determinant as follows,

$$\begin{vmatrix} 1 & x^2 & e^x & y \\ 0 & 2x & e^x & y' \\ 0 & 2 & e^x & y'' \\ 0 & 0 & e^x & y''' \end{vmatrix} = 0, \Rightarrow 1 \cdot \begin{vmatrix} 2x & e^x & y' \\ 2 & e^x & y'' \\ 0 & e^x & y''' \end{vmatrix} = 0$$

$$\Rightarrow 2x(e^x y''' - e^x y'') - 2(e^x y''' - e^x y') = 0, \Rightarrow 2xe^x y''' - 2xe^x y'''$$

$$-2e^x y''' + 2e^x y' = 0, \Rightarrow 2e^x(xy''' - xy'' - y''' + y') = 0,$$

$$\Rightarrow (x-1)y''' - xy'' + y' = 0.$$

Example: Find the general solution of the equation $(2x-3)y'''-(6x-7)y''+4xy'-4y=0$, if the particular solutions $y_1 = e^x, y_2 = e^{2x}$ are known.

Solution: We make the substitution: $y = y_1 z = e^x z$. The derivatives will be:

$$y' = (e^x z)' = e^x z + e^x z' = e^x(z+z'),$$

$$y'' = [e^x(z+z')]' = e^x(z+z') + e^x(z'+z'') = e^x(z+2z'+z''),$$

$$y''' = [e^x(z+2z'+z'')]' = e^x(z+2z'+z'') + e^x(z'+2z''+z''')$$

$$= e^x(z+3z'+3z''+z''').$$

Note that the derivative of the nth order of the product of two functions y,z can be immediately calculated from the Leibniz formula:

$$y^{(n)}(x) = (y_1 z)^{(n)} = \sum_{i=0}^{n} [C_n^i y_1^{(i)} z^{(n-i)}].$$

Substituting the derivatives into the equation and dividing by e^x, we have:

$$(2x-3)\cdot(z+3z'+3z''+z''')-(6x-7)(z+2z'+z'')$$
$$+4x(z+z')-4z=0.$$

After simple transformations, the equation becomes:

$$(2x-3)z +(6x-9)z'+(6x-9)z''+(2x-3)z'''$$
$$-(6x-7)z -(12x-14)z'-(6x-7)z''+ 4xz +4xz'- 4z = 0,$$
$$\Rightarrow (2x-3)z''' - 2z'' -(2x-5)z' = 0.$$

By setting $z' = u$, we obtain a homogeneous linear second-order equation:

$$(2x-3)u'' - 2u' -(2x-5)u=0.$$

Its order can be reduced again by one using the known second particular solution $y_2 = e^{2x}$. The function z^2 corresponds to this solution, so that we can write:

$$y_2 = y_1 z_2, \Rightarrow z_2 = \frac{y_2}{y_1} = \frac{e^{2x}}{e^x} = e^x.$$

From this we obtain a particular solution u_1:

$$u_1 = z_2' = (e^x)' = e^x.$$

Further we act in the same manner. We make the following change:

$$u = u_1 v = e^x v, \Rightarrow u' = e^x(v+v'), \Rightarrow u'' = e^x(v+2v'+v'').$$

We obtain the differential equation for the new variable v:

$$(2x-3)(v+2v'+v'')-2(v+v')-(2x-5)v=0,$$
$$\Rightarrow (2x-3)v+(4x-6)v'+(2x-3)v''- 2v - 2v'- (2x-5)v = 0,$$
$$\Rightarrow (2x-3)v''+(4x-8)v' = 0.$$

Denote $v' = w$. Then we can write:

$$(2x-3)w'+(4x-8)w=0.$$

The last equation is a first-order equation with separable variables.

We find its general solution:

$$(2x-3)\frac{dw}{dx} = -(4x-8)w, \Rightarrow \frac{dw}{w} = -\frac{4x-8}{2x-3}dx,$$

$$\Rightarrow \int \frac{dw}{w} = -\int \frac{4x-8}{2x-3}dx, \Rightarrow \int \frac{dw}{w} = -\int \left(2 - \frac{2}{2x-3}\right)dx,$$

$$\Rightarrow \ln|w| = -2x + \ln|2x-3| + \ln C_1, \Rightarrow \ln|w| = \ln e^{-2x} + \ln|2x-3| + \ln C_1,$$

$$\Rightarrow \ln|w| = \ln(C_1|2x-3|e^{-2x}), \Rightarrow w = C_1(2x-3)e^{-2x}.$$

Now we can restore the function v by integrating the resulting expression for w:

$$v = \int w\,dx = C_1 \int (2x-3)e^{-2x}dx.$$

This integral is calculated by parts:

$$v = C_1 \int (2x-3)e^{-2x}dx = C_1(2x-3)\left(-\frac{1}{2}\right)e^{-2x} - \int 2\left(-\frac{1}{2}\right)e^{-2x}dx$$

$$= C_1\left[\left(-x+\frac{3}{2}\right)e^{-2x} + \int e^{-2x}dx\right] = C_1\left[\left(-x+\frac{3}{2}\right)e^{-2x} - \frac{1}{2}e^{-2x}\right] + C_2$$

$$= -C_1 e^{-2x}\left(x - \frac{3}{2} + \frac{1}{2}\right) + C_2 = -C_1(x-1)e^{-2x} + C_2.$$

Next we find the function u:

$$u = u_1 v = e^x v = e^x[-C_1(x-1)e^{-2x} + C_2] = -C_1(x-1)e^{-x} + C_2 e^x.$$

Performing another integration, we find the function z:

$$z = \int u\,dx = \int [-C_1(x-1)e^{-x} + C_2 e^x]dx$$

$$= -C_1 \int (x-1)e^{-x}dx + C_2 \int e^x dx = -C_1\left[-(x-1)e^{-x} - \int(-e^{-x})dx\right]$$

$$+ C_2 \int e^x dx = -C_1[-(x-1)e^{-x} - e^{-x}] + C_2 e^x + C_3 = C_1 x e^{-x} + C_2 e^x + C_3.$$

Finally, we find the general solution y(x):

$$y = e^x z = e^x(C_1 x e^{-x} + C_2 e^x + C_3) = C_1 x + C_2 e^{2x} + C_3 e^x,$$

where C_1, C_2, C_3 are arbitrary numbers.

Higher Order Linear Non-homogeneous Differential Equations with Constant Coefficients

Higher order linear non-homogeneous differential equations with constant coefficients, have the form:

$$y^{(n)}(x) + a_1 y^{(n-1)}(x) + \cdots + a_{n-1}y'(x) + a_n y(x) = f(x),$$

where $a_1, a_2, \ldots, a_n$ are real or complex numbers, and the right-hand side $f(x)$ is a continuous function on some interval $[a, b]$.

Using the linear differential operator L (D) equal to:

$$L(D) = D^n + a_1 D^{n-1} + \cdots + a_{n-1} D + a_n,$$

the non-homogeneous differential equation can be written as:

$$L(D)y(x) = f(x).$$

The general solution $y(x)$ of the nonhomogeneous equation is the sum of the general solution $y_0(x)$ of the corresponding homogeneous equation and a particular solution $y_1(x)$ of the nonhomogeneous equation:

$$y(x) = y_0(x) + y_1(x).$$

For an arbitrary right side $f(x)$, the general solution of the nonhomogeneous equation can be found using the method of variation of parameters. If the right-hand side is the product of a polynomial and exponential functions, it is more convenient to seek a particular solution by the method of undetermined coefficients.

Method of Variation of Parameters

We assume that the general solution of the homogeneous differential equation of the nth order is known and given by:

$$y_0(x) = C_1 Y_1(x) + C_2 Y_2(x) + \cdots + C_n Y_n(x).$$

According to the method of variation of constants (or Lagrange method), we consider the functions $C_1(x), C_2(x), \ldots, C_n(x)$ instead of the regular numbers $C_1, C_2, \ldots, C_n$. These functions are chosen so that the solution:

$$y = C_1(x)Y_1(x) + C_2(x)Y_2(x) + \cdots + C_n(x)Y_n(x)$$

satisfies the original nonhomogeneous equation.

The derivatives of n unknown functions $C_1(x), C_2(x), \ldots, C_n(x)$ are determined from the system of n equations:

$$\begin{cases} C_1'(x)Y_1(x) + C_2'(x)Y_2(x) + \cdots + C_n'(x)Y_n(x) = 0 \\ C_1'(x)Y_1'(x) + C_2'(x)Y_2'(x) + \cdots + C_n'(x)Y_n'(x) = 0 \\ \ldots\ldots\ldots\ldots\ldots \\ C_1'(x)Y_1^{(n-1)}(x) + C_2'(x)Y_2^{(n-1)}(x) + \cdots + C_n'(x)Y_n^{(n-1)}(x) = f(x). \end{cases}$$

The determinant of this system is the Wronskian of $Y_1, Y_2, \ldots, Y_n$ forming a fundamental system of solutions. By the linear independence of these functions, the determinant is not zero and the system is uniquely solvable. The final expressions for the functions $C_1(x), C_2(x), \ldots, C_n(x)$ can be found by integration.

Method of Undetermined Coefficients

If the right-hand side $f(x)$ of the differential equation is a function of the form:

$$P_n(x)e^{\alpha x} \text{ or } [P_n(x)\cos \beta x + Q_m(x)\sin \beta x]e^{\alpha x},$$

where $P_n(x)$, $Q_m(x)$ are polynomials of degree n and m, respectively, then the method of undetermined coefficients may be used to find a particular solution.

In this case, we seek a particular solution in the form corresponding to the structure of the right-hand side of the equation. For example, if the function has the form:

$$f(x) = P_n(x)e^{\alpha x},$$

the particular solution is given by:

$$y_1(x) = x^s A_n(x)e^{\alpha x},$$

where $A_n(x)$ is a polynomial of the same degree n as $P_n(x)$. The coefficients of the polynomial $A_n(x)$ are determined by direct substitution of the trial solution $y_1(x)$ in the nonhomogeneous differential equation.

In the so-called resonance case, when the number of α in the exponential function coincides with a root of the characteristic equation, an additional factor x^s, where s is the multiplicity of the root, appears in the particular solution. In the *non-resonance case*, we set s=0.

The same algorithm is used when the right-hand side of the equation is given in the form:

$$f(x) = [P_n(x)\cos \beta x + Q_m(x)\sin \beta x]e^{\alpha x}.$$

Here the particular solution has a similar structure and can be written as:

$$y_1(x) = x^s[A_n(x)\cos \beta x + B_n(x)\sin \beta x]e^{\alpha x},$$

where $A_n(x), B_n(x)$ are polynomials of degree n (for $n \geq m$), and the degree s in the additional factor x^s is equal to the multiplicity of the complex root $\alpha \pm \beta i$ in the resonance case (i.e. when the numbers α and β coincide with the complex root of the characteristic equation), and accordingly, s=0 in the non-resonance case.

Superposition Principle

The superposition principle is stated as follows. Let the right-hand side $f(x)$ be the sum of two functions:

$$f(x) = f_1(x) + f_2(x).$$

Suppose that $y_1(x)$ is a solution of the equation:

$$L(D)y(x) = f_1(x),$$

and the function $y_2(x)$ is, accordingly, a solution of the second equation:

$$L(D)y(x) = f_2(x).$$

Then the sum of the functions:

$$y(x) = y_1(x) + y_2(x)$$

will be a solution of the linear non-homogeneous equation:

$$L(D)y(x) = f(x) = f_1(x) + f_2(x).$$

Example: Find the general solution of the differential equation $y''' + 3y'' - 10y' = x - 3$.

Solution: First we find the general solution of the homogeneous equation:

$$y''' + 3y'' - 10y' = 0.$$

Calculate the roots of the characteristic equation:

$$\lambda^3 + 3\lambda^2 - 10\lambda = 0, \Rightarrow \lambda(\lambda^2 + 3\lambda - 10) = 0, \Rightarrow \lambda(\lambda - 2)(\lambda + 5) = 0.$$

Hence,

$$\lambda_1 = 0, \ \lambda_2 = 2, \ \lambda_3 = -5.$$

So the general solution of the homogeneous equation is given by:

$$y_0(x) = C_1 + C_2 e^{2x} + C_3 e^{-5x},$$

where C_1, C_2, C_3 are arbitrary numbers.

The right side of the equation contains only a polynomial. However, if we take into account that $e^0 = 1$, we see that in fact we have the resonance case (in disguised form) as one of the roots of the characteristic equation is also zero: $\lambda_1 = 0$. Therefore, we will seek a particular solution in the form

$$y_1(x) = x(Ax + B) = Ax^2 + Bx.$$

Substitute the derivatives:

$$y_1' = 2Ax + B, \ y_1'' = 2A, \ y_1''' = 0.$$

into the nonhomogeneous equation and determine the coefficients A, B:

$$0 + 3 \cdot 2A - 10(2Ax + B) = x - 3, \Rightarrow 6A - 20Ax - 10B = x - 3,$$

$$\Rightarrow \begin{cases} -20A = 1 \\ 6A - 10B = -3 \end{cases}, \Rightarrow \begin{cases} A = -\dfrac{1}{20} \\ B = \dfrac{27}{100} \end{cases}, \Rightarrow \begin{cases} A = -\dfrac{5}{100} \\ B = \dfrac{27}{100} \end{cases}.$$

The particular solution y_1 is written as:

$$y_1(x) = x\left(-\frac{5}{100}x + \frac{27}{100}\right) = \frac{x}{100}(27 - 5x).$$

Thus, the general solution of non-homogeneous differential equation is given by:

$$y(x) = y_0(x) + y_1(x) = C_1 + C_2 e^{2x} + C_3 e^{-5x} + \frac{x}{100}(27 - 5x).$$

Example: Solve the differential equation $y''' - y' = \sin 3x$.

Solution: We construct the general solution of the homogeneous equation:

$$y''' - y' = 0.$$

The roots of the characteristic equation are:

$$\lambda^3 - \lambda = 0, \Rightarrow \lambda(\lambda^2 - 1) = 0, \Rightarrow \lambda(\lambda - 1)(\lambda + 1) = 0, \Rightarrow \lambda_1 = 0, \lambda_2 = 1,$$
$$\lambda_3 = -1.$$

Consequently, the general solution of the homogeneous equation can be written as:

$$y_0(x) = C_1 + C_2 e^x + C_3 e^{-x},$$

where C_1, C_2, C_3 are arbitrary numbers.

Based on the structure of the right-hand side, we seek a particular solution in the form of trial function:

$$y(x) = A \sin 3x + B \cos 3x.$$

The derivatives of this function are as follows:

$$y_1' = 3A \cos 3x - 3B \sin 3x,$$
$$y_1'' = -9A \sin 3x - 9B \cos 3x,$$
$$y_1''' = -27A \cos 3x + 27B \sin 3x.$$

Substituting these derivatives into the equation, we obtain:

$$-27A \cos 3x + 27B \sin 3x - 3A \cos 3x + 3B \sin 3x = \sin 3x,$$

$$\Rightarrow -30A \cos 3x + 30B \sin 3x = \sin 3x, \Rightarrow \begin{cases} -30A = 0 \\ 30B = 1 \end{cases}, \Rightarrow \begin{cases} A = 0 \\ B = \dfrac{1}{30} \end{cases}$$

Thus, a particular solution can be written as:

$$y_1(x) = \frac{1}{30} \cos 3x.$$

Accordingly, the general solution of the nonhomogeneous equation is described by:

$$y(x) = y_0(x) + y_1(x) = C_1 + C_2 e^x + C_3 e^{-x} + \frac{1}{30} \cos 3x.$$

Example: Solve the differential equation $y^{IV} - y = 2\cos x$.

Solution: We first consider the homogeneous equation,

$$y^{IV} - y = 0$$

and construct its general solution. The characteristic equation:

$$\lambda^4 - 1 = 0$$

has the following roots:

$$(\lambda^2 - 1)(\lambda^2 + 1) = 0, \Rightarrow (\lambda - 1)(\lambda + 1)(\lambda^2 + 1) = 0, \Rightarrow \lambda_1 = 1, \lambda_2 = -1,$$
$$\lambda_{3,4} = \pm i.$$

Consequently, the general solution of the homogeneous equation has the form:

$$y_0(x) = C_1 e^x + C_2 e^{-x} + C_3 \cos x + C_4 \sin x,$$

where $C_1, \ldots, C_4$ are arbitrary numbers.

Now we find a particular solution of the nonhomogeneous equation. Here we have the resonance case, since the expression in the right side corresponds to one of the roots of the characteristic equation. Hence, we seek a particular solution in the form:

$$y_1(x) = x(A\cos x + B\sin x).$$

The derivatives of this function are:

$$y_1' = A\cos x + B\sin x + x(-A\sin x + B\cos x),$$
$$y_1'' = -A\sin x + B\cos x + (-A\sin x + B\cos x) + x(-A\cos x - B\sin x)$$
$$= -2A\sin x + 2B\cos x - x(A\cos x + B\sin x),$$
$$y_1''' = -2A\cos x - 2B\sin x - (A\cos x + B\sin x) - x(-A\sin x + B\cos x)$$
$$= -3A\cos x - 3B\sin x + x(A\sin x - B\cos x),$$
$$y^{IV} = 3A\sin x - 3B\cos x + (A\sin x - B\cos x) + x(A\cos x + B\sin x)$$
$$= 4A\sin x - 4B\cos x + x(A\cos x + B\sin x).$$

Substitute the derivatives in the nonhomogeneous equation and determine the coefficients A, B:

$$4A\sin x - 4B\cos x + x\underline{(A\cos x + B\sin x)} - x\underline{(A\cos x + B\sin x)} = 2\cos x,$$

$$\Rightarrow \begin{cases} 4A = 0 \\ -4B = 2 \end{cases}, \Rightarrow \begin{cases} A = 0 \\ B = -\dfrac{1}{2} \end{cases}.$$

Thus, a particular solution is expressed as:

$$y_1(x) = -\frac{x}{2}\sin x.$$

Then the general solution of the original nonhomogeneous equation can be written as:

$$y(x) = y_0(x) + y_1(x) = C_1 e^x + C_2 e^{-x} + C_3 \cos x + C_4 \sin x - \frac{x}{2} \sin x.$$

Example: Solve the equation $y^{IV} + y''' - 3y'' - 5y' - 2y = e^{2x} - e^{-x}$.

Solution: First we find the general solution of the homogeneous equation,

$$y^{IV} + y''' - 3y'' - 5y' - 2y = 0.$$

Write the characteristic equation and find its roots:

$$\lambda^4 + \lambda^3 - 3\lambda^2 - 5\lambda - 2 = 0, \Rightarrow \lambda^4 - 2\lambda^3 + 3\lambda^3 - 6\lambda^2 + 3\lambda^2 - 6\lambda + \lambda - 2 = 0,$$

$$\Rightarrow \lambda^3(\lambda - 2) + 3\lambda^2(\lambda - 2) + 3\lambda(\lambda - 2) + \lambda - 2 = 0,$$

$$\Rightarrow (\lambda^3 + 3\lambda^2 + 3\lambda + 1) \cdot (\lambda - 2) = 0, \Rightarrow (\lambda + 1)^3 (\lambda - 2) = 0.$$

It is seen that the equation has two roots:

$$\lambda_1 = -1, \ \lambda_2 = 2,$$

and the multiplicity of the first root is 3.

Then the general solution of the homogeneous equation can be written as:

$$y_0(x) = (C_1 + C_2 x + C_3 x^2)e^{-x} + C_4 e^{2x},$$

where $C_1, \ldots, C_4$ are as usual arbitrary numbers.

We now construct a particular solution of the nonhomogeneous equation. Using the superposition principle, it is convenient to consider two nonhomogeneous equations of the form:

$$y^{IV} + y''' - 3y'' - 5y' - 2y = e^{2x};$$

$$y^{IV} + y''' - 3y'' - 5y' - 2y = -e^{-x}.$$

The sum of the right sides of these equations corresponds to the right side of the original nonhomogeneous equation.

We have the resonance cases in both equations. In the first equation the number 2 in the exponential function coincides with the root $\lambda_2 = 2$ of multiplicity 2. In the second equation the number -1 in the exponential function coincides with another root $\lambda_1 = -1$, the multiplicity of which is equal to 3. With this in mind, we seek particular solutions y_1, y_2, respectively, for equations above in the form:

$$y_1 = Axe^{2x}, \ y_2 = Bx^3 e^{-x}.$$

The derivatives for the trial solution y_1 have the form:

$$y_1' = A(e^{2x} + 2xe^{2x}) = A(2x+1)e^{2x},$$

$$y_1'' = A[2e^{2x} + (4x+2)e^{2x}] = A(4x+4)e^{2x},$$

$$y_1''' = A[4e^{2x} + (8x+8)e^{2x}] = A(8x+12)e^{2x},$$

$$y_1^{IV} = A[8e^{2x} + (16x+24)e^{2x}] = A(16x+32)e^{2x}.$$

Substituting this into the first equation, we find the coefficient A:

$$A(16x+32)e^{2x} + A(8x+12)e^{2x} - 3A(4x+4)e^{2x} - 5A(2x+1)e^{2x} - 2Axe^{2x}$$
$$= e^{2x},$$
$$\Rightarrow A(\cancel{16x} + \cancel{8x} - \cancel{12x} - \cancel{10x} - \cancel{2x})e^{2x} + A(32 + \cancel{12} - \cancel{12} - 5)e^{2x} = e^{2x},$$
$$\Rightarrow 27A = 1, \Rightarrow A = \frac{1}{27}.$$

Therefore, the particular solution y_1 is given by:

$$y_1(x) = \frac{x}{27}e^{2x}.$$

Similarly, we find the particular solution y_2. The derivatives of the trial function y_2 are:

$$y_2' = B(3x^2e^{-x} - x^3e^{-x}) = B(-x^3 + 3x^2)e^{-x},$$
$$y_2'' = B[(-3x^2 + 6x)e^{-x} - (-x^3 + 3x^2)e^{-x}] = B(x^3 - 6x^2 + 6x)e^{-x},$$
$$y_2''' = B[(3x^2 - 12x + 6)e^{-x} - (x^3 - 6x^2 + 6x)e^{-x}] = B(-x^3 + 9x^2 - 18x + 6)e^{-x},$$
$$y_2^{IV} = B[(-3x^2 + 18x - 18)e^{-x} - (-x^3 + 9x^2 - 18x + 6)e^{-x}] = B(x^3 - 12x^2 + 36x - 24)e^{-x}.$$

Substituting these derivatives into the second equation, we calculate the coefficient B:

$$B(x^3 - 12x^2 + 36x - 24)e^{-x} + B(-x^3 + 9x^2 - 18x + 6)e^{-x}$$
$$-3B(x^3 - 6x^2 + 6x)e^{-x} - 5B(-x^3 + 3x^2)e^{-x} - 2Bx^3e^{-x} = -e^{-x},$$
$$\Rightarrow B(\cancel{x^3} - \cancel{x^3} - \cancel{3x^3} + \cancel{5x^3} - \cancel{2x^3})e^{-x} + B(-\cancel{12x^2} + 9x^2$$
$$+\cancel{18x^2} - \cancel{15x^2})e^{-x} + B(\cancel{36x} - \cancel{18x} - \cancel{18x})e^{-x} + B(-24 + 6)e^{-x} = -e^{-x},$$
$$\Rightarrow -18B = -1, \Rightarrow B = \frac{1}{18}.$$

We obtain the solution y_2 as follows:

$$y_2(x) = \frac{x^3}{18}e^{-x}.$$

In accordance with the principle of superposition, a particular solution of the original nonhomogeneous equation is represented as:

$$y_p = y_1(x) + y_2(x) = \frac{x}{27}e^{2x} + \frac{x^3}{18}e^{-x}.$$

Finally, the general solution is given by:

$$y(x) = (C_1 + C_2x + C_3x^2)e^{-x} + C_4e^{2x} + \frac{x}{27}e^{2x} + \frac{x^3}{18}e^{-x}$$
$$= (C_1 + C_2x + C_3x^2 + \frac{x^3}{18})e^{-x} + (C_4 + \frac{x}{27})e^{2x}.$$

Example: Find the general solution of the equation $y''' + y' = \dfrac{1}{\cos x}$ using the method of variation of constants.

Solution: First we solve the corresponding homogeneous equation,

$$y''' + y' = 0.$$

The roots of its characteristic equation are:

$$\lambda^3 + \lambda = 0, \Rightarrow \lambda(\lambda^2 + 1) = 0, \Rightarrow \lambda_1 = 0, \lambda_{2,3} = \pm i.$$

Consequently, the general solution of the homogeneous equation has the form:

$$y_0(x) = C_1 + C_2 \cos x + C_3 \sin x,$$

where C_1, C_2, C_3 are arbitrary numbers.

According to the method of variation of constants, we will consider the functions $C_1(x), C_2(x), C_3(x)$ instead of the numbers C_1, C_2, C_3 to construct the general solution of the non-homogeneous equation. These functions will satisfy the non-homogeneous equation, provided

$$\begin{cases} C_1' Y_1 + C_2' Y_2 + C_3' Y_3 = 0 \\ C_1' Y_1' + C_2' Y_2' + C_3' Y_3' = 0 \\ C_1' Y_1'' + C_2' Y_2'' + C_3' Y_3'' = \dfrac{1}{\cos x} \end{cases}$$

Here the functions Y_1, Y_2, Y_3 are the fundamental system of solutions. They were found in the solution of the homogeneous equation:

$$Y_1 = 1, \, Y_2 = \cos x, \, Y_3 = \sin x.$$

Then the system of equations takes the form:

$$\begin{cases} C_1' \cdot 1 + C_2' \cos x + C_3' \sin x = 0 \\ C_1' \cdot 0 + C_2'(-\sin x) + C_3' \cos x = 0 \\ C_1' \cdot 0 + C_2'(-\cos x) + C_3'(-\sin x) = \dfrac{1}{\cos x} \end{cases}$$

$$\Rightarrow \begin{cases} C_1' + C_2' \cos x + C_3' \sin x = 0 \\ -C_2' \sin x + C_3' \cos x = 0 \\ -C_2' \cos x - C_3' \sin x = \dfrac{1}{\cos x} \end{cases}$$

The main determinant (Wronskian) is:

$$W = \begin{vmatrix} 1 & \cos x & \sin x \\ 0 & -\sin x & \cos x \\ 0 & -\cos x & -\sin x \end{vmatrix} = 1 \cdot \begin{vmatrix} -\sin x & \cos x \\ -\cos x & -\sin x \end{vmatrix} = \sin^2 x + \cos^2 x = 1.$$

We find expressions for the derivatives C_1', C_2', C_3' calculating the other three determinants:

$$\Delta_1 = \begin{vmatrix} 0 & \cos x & \sin x \\ 0 & -\sin x & \cos x \\ \dfrac{1}{\cos x} & -\cos x & -\sin x \end{vmatrix} = \frac{1}{\cos x} \begin{vmatrix} \cos x & \sin x \\ -\sin x & \cos x \end{vmatrix} = \frac{1}{\cos x}(\cos^2 x + \sin^2 x)$$

$$= \frac{1}{\cos x},$$

$$\Delta_2 = \begin{vmatrix} 1 & 0 & \sin x \\ 0 & 0 & \cos x \\ 0 & \dfrac{1}{\cos x} & -\sin x \end{vmatrix} = 1 \cdot \begin{vmatrix} 0 & \cos x \\ \dfrac{1}{\cos x} & -\sin x \end{vmatrix} = -\frac{1}{\cos x} \cdot \cos x = -1,$$

$$\Delta_3 = \begin{vmatrix} 1 & \cos x & 0 \\ 0 & -\sin x & 0 \\ 0 & -\cos x & \dfrac{1}{\cos x} \end{vmatrix} = 1 \cdot \begin{vmatrix} -\sin x & 0 \\ -\cos x & \dfrac{1}{\cos x} \end{vmatrix} = -\sin x \cdot \frac{1}{\cos x} = -\tan x.$$

Consequently, the derivatives C_1', C_2', C_3' are given by:

$$C_1' = \frac{\Delta_1}{W} = \frac{1}{\cos x}, \; C_2' = \frac{\Delta_2}{W} = -1, \; C_3' = \frac{\Delta_3}{W} = -\tan x.$$

The integrals of these functions are tabulated, so that we can immediately write:

$$C_1(x) = \int \frac{dx}{\cos x} = \ln\left|\tan\left(\frac{x}{2} + \frac{\pi}{4}\right)\right| + A_1,$$

$$C_2(x) = \int (-1)dx = -x + A_2,$$

$$C_3(x) = \int (-\tan x)dx = \ln|\cos x| + A_3,$$

where A_1, A_2, A_3, are constants of integration.

Substituting this into the general solution, we find the answer in the following form:

$$y(x) = C_1(x) + C_2(x)\cos x + C_3(x)\sin x = \ln\left|\tan\left(\frac{x}{2} + \frac{\pi}{4}\right)\right| + A_1$$

$$+ (-x + A_2)\cos x + (\ln|\cos x| + A_3)\sin x = A_1 + A_2\cos x + A_3\sin x$$

$$+ \ln\left|\tan\left(\frac{x}{2} + \frac{\pi}{4}\right)\right| - x\cos x + \sin x \ln|\cos x|.$$

System of Simultaneous Linear Differential Equations with Constant Coefficients

We will discuss systems of simultaneous linear differential equations which contain a single independent variable and two or more dependent variables. In general, the number of equations will be equal to the number of dependent variables i.e. if there are n dependent variables there will be n equations.

Examples of Systems

Example: The following two-equation system where x and y are dependent variables and t is the independent variable:

$$2\frac{dx}{dt} + \frac{dy}{dt} - 4x - y = e'$$

$$\frac{dx}{dt} + 3x + y = 0$$

or, equivalently,

$$2(D-2)x + (D-1)y = e^t$$

$$(D+3)x + y = 0$$

where D = d/dt.

Example: The following three-equation system where x, y and z are dependent variables and t is the independent variable:

$$\frac{dx}{dt} + \frac{dy}{dt} + y = 1$$

$$\frac{dx}{dt} - \frac{dz}{dt} + 2x + z = 1$$

$$\frac{dy}{dt} + \frac{dz}{dt} + y + 2z = 0$$

or, equivalently,

$$Dx + (D+1)y = 1$$
$$(D+2)x - (D-1)z = 1$$
$$(D+1)y + (D+2)z = 0$$

where $D = d/dt$.

Each equation in a system will be assumed to have constant coefficients and be of the general form:

$$f(D)x + g(D)y + \dots\dots + h(D)u = G(t)$$

where x, y, ..., u are the dependent variables, t is the independent variable.

A system of n equations in the n dependent variables x, y, ..., u and independent variable t will have a solution consisting of n functions:

$$x = x(t)$$
$$y = y(t)$$
$$\dots\dots\dots\dots$$
$$\dots\dots\dots\dots$$
$$u = u(t)$$

The method used to solve a system of n equations in n variables is analogous to the procedure used to solve a system of n linear equations in n unknowns in algebra. In algebra, we solve a system of n equations in n unknowns by eliminating unknowns between equations until we obtain an equation containing a single unknown, from which we deduce the value of the unknown. Then we substitute the value of that unknown into other equations to obtain the values of other unknowns. In solving systems of linear differential equations we go through the same type process to obtain an equation containing a single dependent variable. The equation in this single dependent variable will be a linear differential equation with constant coefficients. We then solve this equation, using methods for solving such equations, to obtain an expression for that dependent variable. We then substitute the expression for that variable into another equation to obtain an expression for another variable. As with the algebraic problem, we can also employ determinants.

Theorem: The number of constants in the general solution of a system of equations must equal the sum of the orders of the equations.

Example: Solve the linear system,

$$y'' - y + 5v' = x$$
$$2y' - v'' + 4v = 2$$

Method 1

Step 1: Put the equations into operator form.

$$(D^2 - 1)y + 5Dv = x$$

$$2Dy - \left(D^2 - 4\right)v = 2$$

Step 2: Obtain an equation in v alone by eliminating y.

Multiplying equations $\left(D^2 - 1\right)y + 5Dv = x$ through by $2D$ and $2Dy - \left(D^2 - 4\right)v = 2$ through by $\left(D^2 - 1\right)$ and then subtracting one from the other we obtain:

$$\left[10D^2 + \left(D^2 - 1\right)\left(D^2 - 4\right)\right]v = 2Dx - \left(D^2 - 1\right)2$$

or,

$$\left(D^4 + 5D^2 + 4\right)v = 4$$

When we speak of "multiplying" by an operator we really mean "operate on" by the operator.

Step 3: Solve equation $\left(D^4 + 5D^2 + 4\right)v = 4$ for v.

Using methods for solving linear differential equations with constant coefficients we find the solution as:

$$v = 1 + a_1 \cos x + a_2 \sin x + a_3 \cos 2x + a_4 \sin 2x$$

Step 4: Obtain an equation in y alone.

We could, if we wished, find an equation in y using the same method as we used in Step 2. However, for the purpose of demonstration, we will find y using determinants. Using Cramer's Rule on the system of equations $(D^2 - 1)y + 5Dv = x$ and $2Dy - (D^2 - 4)v = 2$ we get,

$$\begin{vmatrix} D^2 - 1 & 5D \\ 2D & -(D^2 - 4) \end{vmatrix} y = \begin{vmatrix} x & 5D \\ 2 & -(D^2 - 4) \end{vmatrix}$$

or

$$\left[-\left(D^2 - 1\right)\left(D^2 - 4\right) - 10D^2\right]y = 4x$$

or

$$\left(D^4 + 5D^2 + 4\right)y = -4x$$

We must take care to interpret the right member of equation $\begin{vmatrix} D^2 - 1 & 5D \\ 2D & -(D^2 - 4) \end{vmatrix} y = \begin{vmatrix} x & 5D \\ 2 & -(D^2 - 4) \end{vmatrix}$ correctly.

It is interpreted as:

$$-\left(D^2 - 4\right)(x) - 5D(2)$$

Step 5: Solve equation $\left(D^4 + 5D^2 + 4\right)y = -4x$ for y. The solution is:

$$y = -x + b_1 \cos x + b_2 \sin x + b_3 \cos 2x + b_4 \sin 2x$$

Now the number of constants in the solution must be four since the number of constants must equal the sum of the orders of the equations in the original system. Since equation $y'' - y + 5v' = x$

is of order $2y' - v'' + 4v = 2$ the sum of the orders is four. Now equations $v = 1 + a_1 \cos x + a_2 \sin x + a_3 \cos 2x + a_4 \sin 2x$ and $y = -x + b_1 \cos x + b_2 \sin x + b_3 \cos 2x + b_4 \sin 2x$ must satisfy the original equations rather than just equations $(D^4 + 5D^2 + 4)v = 4$ and $(D^4 + 5D^2 + 4)y = -4x$ which resulted from the original ones after some manipulations were performed (the technique we used introduced extra constants).

Step 6: Substitute equations $v = 1 + a_1 \cos x + a_2 \sin x + a_3 \cos 2x + a_4 \sin 2x$ and $y = -x + b_1 \cos x + b_2 \sin x + b_3 \cos 2x + b_4 \sin 2x$ into equation $y'' - y + 5v' = x$.

We get,

$$x - 2b_1 \cos x - 2b_2 \sin x - 5b_3 \cos 2x - 5b_4 \sin 2x - 5a_1$$
$$\sin x + 5a_2 \cos x - 10a_3 \sin 2x + 10a_4 \cos 2x \equiv x$$

which is an identity in x.

Step 7: Equate coefficients of corresponding terms in above equation to obtain the final coefficients.

The fact that equation $x - 2b_1 \cos x - 2b_2 \sin x - 5b_3 \cos 2x - 5b_4 \sin 2x - 5a_1 \sin x + 5a_2 \cos x - 10a_3 \sin 2x + 10a_4 \cos 2x \equiv x$ is an identity in x demands that:

$$-2b_1 + 5a_2 = 0$$
$$-2b_2 - 5a_1 = 0$$
$$-5b_3 + 10a_4 = 0$$
$$-5b_4 - 10a_3 = 0$$

If we were to go through steps 6 and 7 for equation $2y' - v'' + 4v = 2$ we would get equivalent results.

Step 8: Write down the general solution. It is,

$$v = 1 + a_1 \cos x + a_2 \sin x + a_3 \cos 2x + a_4 \sin 2x$$

$$y = -x + \frac{5}{2}a_2 \cos x \frac{5}{2}a_1 \sin x + 2a_4 \cos 2x - 2a_3 \sin 2x$$

Method 2

We shall now solve the same problem by a different method which is simpler and easier. It is the same as Method 1 up through Step 3, but finds y in a different way.

Step 1: Put the equations into operator form.

$$\left(D^2 - 1\right)y + 5Dv = x$$
$$2Dy - \left(D^2 - 4\right)v = 2$$

Step 2: Obtain an equation in v alone by eliminating y.

Multiplying equation (D² - 1)y + 5Dv = x through by 2D and 2Dy - (D² - 4)v = 2 through by (D² - 1) and then subtracting one from the other we obtain:

$$\left[10D^2 + \left(D^2 - 1\right)\left(D^2 - 4\right)\right]v = 2Dx - \left(D^2 - 1\right)2$$

or

$$\left(D^4 + 5D^2 + 4\right)v = 4$$

Step 3: Solve $\left(D^4 + 5D^2 + 4\right)v = 4$ for v.

Using methods for solving linear differential equations with constant coefficients we find the solution as:

$$v = 1 + a_1 \cos x + a_2 \sin x + a_3 \cos 2x + a_4 \sin 2x$$

Step 4: Find an equation giving y in terms of v. We accomplish this by eliminating from the system of equation $\left(D^2 - 1\right)y + 5Dv = x$ and $2Dy - \left(D^2 - 4\right)v = 2$ those terms which involve derivatives of y.

Multiplying equation (D² - 1)y + 5Dv = x by 2 and 2Dy - (D² - 4)v = 2 by D we get:

$$\left(2D^2 - 2\right)y + 10Dv = 2x$$
$$2D^2y - \left(D^3 - 4D\right)v = 0$$

Subtracting above equations we get:

$$2y - D\ v - 6Dv = -2x$$

or

$$y = -x + \frac{1}{2}D^3v + 3Dv$$

Step 5: Substitute equation $v = 1 + a_1 \cos x + a_2 \sin x + a_3 \cos 2x + a_4 \sin 2x$ into $y = -x + \frac{1}{2}D^3v + 3Dv$ to obtain the expression for y. It is:

$$y = -x + \frac{5}{2}a_2 \cos x - \frac{5}{2}a_1 \sin x + 2a_4 \cos 2x - 2a_3 \sin 2x$$

Step 6: Write down the general solution. It is:

$$v = 1 + a_1 \cos x + a_2 \sin x + a_3 \cos 2x + a_4 \sin 2x$$

$$y = -x + \frac{5}{2}a_2 \cos x - \frac{5}{2}a_1 \sin x + 2a_4 \cos 2x - 2a_3 \sin 2x$$

We have given both methods to illustrate the superiority of the second method. The second method avoids introducing extra constants. Thus it is best to use the second method whenever possible.

We will now give another example to make a point. The point to be made is that the use of differential operators is not necessary to solving systems of linear differential equations. They make the work much easier but what can be done with them can also be done without them.

Example: Solve the following system.

$$2\frac{dx}{dt} + \frac{dy}{dt} - 4x - y = e^t$$

$$\frac{dx}{dt} + 3x + y = 0$$

Solution: Differentiating equation $\frac{dx}{dt} + 3x + y = 0$ we get,

$$\frac{d^2x}{dt^2} + 3\frac{dx}{dt} + \frac{dy}{dt} = 0$$

Multiplying equation $2\frac{dx}{dt} + \frac{dy}{dt} - 4x - y = e^t$ by -1, $\frac{dx}{dt} + 3x + y = 0$ by -1, 3) by 1, and adding we get:

$$\frac{d^2x}{dt^2} + x = -e^t$$

Using methods for solving such equations we get as a solution:

$$x = c_1 \cos t + c_2 \sin t - \frac{1}{2}e^t$$

Substituting equation $x = c_1 \cos t + c_2 \sin t - \frac{1}{2}e^t$ into $\frac{dx}{dt} + 3x + y = 0$ we get:

$$y = -\frac{dx}{dt} - 3x = -(-c_1 \sin t + c_2 \cos t - \frac{1}{2}e^t) - 3(c_1 \cos t + c_2 \sin t - \frac{1}{2}e^t)$$

$$= (c_1 - 3c_2)\sin t - (3c_1 + c_2)\cos t + 2e^t$$

Thus the general solution is:

$$x = c_1 \cos t + c_2 \sin t - \frac{1}{2}e^t$$

$$y = (c_1 - 3c_2)\sin t - (3c_1 + c_2)\cos t + 2e^t$$

References

- Second-order-linear-homogeneous-differential-equations-variable-coefficients: math24.net, Retrieved 24 January, 2019

- Nonhomogeneous-Linear-Equations, Second-Order-Differential-Equations: math.libretexts.org, Retrieved 20 June, 2019

- Undetermined-Coefficients: tutorial.math.lamar.edu, Retrieved 08 May, 2019

- Variation-of-Parameters: tutorial.math.lamar.edu, Retrieved 15 April, 2019

- Higher-order-linear-homogeneous-differential-equations-constant-coefficients: math24.net, Retrieved 18 August, 2019

Partial Differential Equations

Partial differentiation equations comprise of an unknown variable and their partial derivatives. This kind of equations can be used in different phenomena like sound, heat, diffusion, fluid dynamics, gravitation, etc. All the aspects related to partial differential equations have been carefully analyzed in this chapter.

A partial differential equation (PDE) is an equation involving functions and their partial derivatives; for example, the wave equation:

$$\frac{\partial^2 \psi}{\partial x^2} + \frac{\partial^2 \psi}{\partial y^2} + \frac{\partial^2 \psi}{\partial z^2} = \frac{1}{v^2}\frac{\partial^2 \psi}{\partial t^2}.$$

Some partial differential equations can be solved exactly in the Wolfram Language using DSolve [eqns, y, {x1, x2}] and numerically using NDSolve [eqns, y, {x, xmin, xmax}, {t, tmin, tmax}].

In general, partial differential equations are much more difficult to solve analytically than are ordinary differential equations. They may sometimes be solved using a Bäcklund transformation, characteristics, Green's function, integral transform, Lax pair, separation of variables, or when all else fails (which it frequently does) numerical methods such as finite differences.

Fortunately, partial differential equations of second-order are often amenable to analytical solution. Such PDEs are of the form:

$$Au_{xx} + 2Bu_{xy} + Cu_{yy} + Du_x + Eu_y + F = 0.$$

Linear second-order PDEs are then classified according to the properties of the matrix:

$$Z = \begin{vmatrix} A & B \\ B & C \end{vmatrix}$$

as elliptic, hyperbolic, or parabolic.

If Z is a positive definite matrix, i.e., $\det(Z) > 0$, the PDE is said to be elliptic. Laplace's equation and Poisson's equation are examples. Boundary conditions are used to give the constraint $u(x,y) = g(x,y)$ on $\partial\Omega$ where:

$$u_{xx} + u_{yy} = f(u_x, u_y, u, x, y)$$

holds in Ω.

If $\det(Z) < 0$, the PDE is said to be hyperbolic. The wave equation is an example of a hyperbolic partial differential equation. Initial-boundary conditions are used to give:

$$u(x,y,t) = g(x,y,t) \text{ for } x \in \partial\Omega, t > 0$$

$$u(x,y,0) = v_0(x,y) \, \text{in} \, \Omega$$
$$u_t(x,y,0) = v_1(x,y) \, \text{in} \, \Omega,$$

where,

$$u_{xy} = f(u_x, u_t, x, y)$$

holds in Ω.

If $\det(Z) = 0$, the PDE is said to be parabolic. The heat conduction equation and other diffusion equations are examples. Initial-boundary conditions are used to give:

$$u(x,t) = g(x,t) \, \text{for} \, x \in \partial\Omega, t > 0$$
$$u(x,0) = v(x) \, \text{for} \, x \in \Omega,$$

where

$$u_{xx} = f(u_x, u_y u, x, y)$$

holds in Ω.

Order of a PDE

The order of a PDE is determined by the highest derivative in the equation. For example,

$$\frac{\partial u}{\partial t} - \frac{u}{\partial u}\partial x = \frac{\partial^2 u}{\partial x^2}\frac{\partial^2 u}{\partial x_1^2} + \frac{\partial^2 u}{\partial x_2^2} + \frac{\partial^2 u}{\partial x_3^2} = 0 \, \text{is a second- or PDE.}$$

$$\frac{\partial^4 u}{\partial x_1^4} + \frac{\partial^2 u}{\partial x_2^2} - u = 0 \, \text{is a fourth-order PDE.}$$

$$\left(\frac{\partial u}{\partial x^1}\right) + \frac{\partial u}{\partial x_2} + u^4 = 0 \, \text{is a first-order PDE.}$$

Homogeneous PDEs

Let L be a linear operator. Then a linear partial differential equation can be written in the form:

$$L(u) = f(x_1, x_2, x_3, t).$$

If $f(x_1, x_2, x_3, t) = 0$, the PDE is called *homogeneous*. For example,

$$\frac{\partial u}{\partial t} + \frac{\partial u}{\partial x_1} + \frac{\partial u}{\partial x_2} + \frac{\partial u}{\partial x_3} = 0 \, \text{is homogeneous.}$$

$$\frac{\partial u}{\partial t} + \frac{\partial u}{\partial x_1} + \frac{\partial u}{\partial x_2} + \frac{\partial u}{\partial x_3} = x_1 + x_2 \, \text{is nonhomogeneous.}$$

First Order Partial Differential Equations

A partial differential equation of order one in its most general form is an equation of the form:

$$F(\vec{x}, u, \nabla u) = 0,$$

where the unknown is the function $u = u(\vec{x}) = u(x_1, ..., x_n)$ of n real variables. Here, we will not consider problems of such generality but will focus instead on a smaller class of problems. For example, the equation $F(\vec{x}, u, \nabla u) = 0,$ is said to be a quasilinear equation in two variables if it is of the form:

$$a(t, x, u)\partial_t u(x, t) + b(t, x, u)\partial_x u(t, x) = f(t, x, u),$$

i.e., the equation is linear in the derivatives $\partial_t u$ and $\partial_x u$ but is nonlinear in u. If $f(t, x, u) = 0$, the equation is said to be homogeneous. In order to make the notation more convenient later, we are choosing to call the independent variables, t and x.

Suppose $u = u(x, t)$t is a smooth solution of equation $a(t, x, u)\partial_i u(x, t) + b(t, x, u)\partial_x u(t, x) = f(t, x, u),$ and let:

$$S = \{(t, x, u) \in R^3 : u = u(x, t)\}.$$

Then S is said to be a solution surface for equation $a(t, x, u)\partial_i u(x, t) + b(t, x, u)\partial_x u(t, x) = f(t, x, u).$

The smoothness of the solution u means that S has a tangent plane at each point $(t, x, u) \in S$. The normal vector $\vec{n}$ to the tangent plane has the direction numbers $(\partial_t u, \partial_x u, -1)$; i.e., $u(x, t) - u = 0$ is the equation of S and $\partial_t u dt + \partial_x u dx - du = 0$ is the equation of the tangent plane.

Now consider a curve $C = \{t = t(s), \ x = x(s), \ u = u(s), s \in 1\}$ in 3-space defined as a solution curve for the system:

$$\frac{dt}{ds} = a(t, x, u), \quad \frac{dx}{ds} = b(t, x, u), \quad \frac{du}{ds} = f(t, x, u),$$

If $\vec{T}$ denotes a vector tangent to Cat (t, x, u) then the direction numbers of $\vec{T}$ must be (a, b, f). But then a quasilinear equation implies that $\vec{T} \perp \vec{n}$, which is to say, $\vec{T}$ lies in the tangent plane to the surface S. But if T lies in the tangent plane, then C must lie in S. Evidently, solution curves of a quasilinear equation lie in the solution surface S associated with this equation. Such curves are called characteristic curves for a quasilinear equation. Note that if C is a solution curve for equation $\frac{dt}{ds} = a(t, x, u), \quad \frac{dx}{ds} = b(t, x, u), \quad \frac{du}{ds} = f(t, x, u),$ then:

$$a(t, x, u)\partial_t u(x, t) + b(t, x, u)\partial_x u(x, t, u) = \partial_t u(x, t)\frac{dt}{ds} + \partial_x u(x, t)\frac{dx}{ds} = \frac{du}{ds} = f(t, x, u),$$

so the pde reduces to an ode along C. In general, the equations for C must be solved as a system. We will recall now some notions from differential geometry that will clarify the procedure for solving the system.

Integral Curves for Vector Fields

A vector valued function, $\vec{V} = (P(t,x,u), Q(t,x,u), R(t,x,u))$ is called a vector field if P,Q, R are all smooth functions and if $P^2 + Q^2 + R^2$ is never zero. A space curve, $C = \{t = t(s), x = x(s), u = u(s), s \in I\}$ is said to be an integral curve or trajectory for $\vec{V}$ if $\vec{V}$ is tangent to C at every point; i.e., if:

$$\frac{dt}{ds} = P(t,x,u), \quad \frac{dx}{ds} = Q(t,x,u), \quad \frac{du}{ds} = R(t,x,u),$$

or, equivalently $\quad \dfrac{dt}{P} = \dfrac{dx}{Q} = \dfrac{du}{R}$

A function $\phi = \phi(t,x,u)$ is said to be a first integral for the vector field $\vec{V} = (P,Q,R)$ if:

$$P(t,x,u)\partial_t\phi + Q(t,x,u)\partial_x\phi + R(t,x,u)\partial_u\phi = 0.$$

The trajectories C for $\vec{V}$ will be found by representing C as the intersection of level surfaces of first integrals. The level surfaces:

$$S_j = \{(t,x,u) : \phi_j(t,x,u) = C_j\} \quad j = 1,2$$

intersect transversally at each point if their normals, $\vec{n}_1$ and $\vec{n}_2$ are never parallel. This situation occurs if ϕ_1 and ϕ_2 are such that the expression $\nabla\phi_1 \times \nabla\phi_2$ is different from zero at each point. In this case the functions ϕ_1 and ϕ_2 are said to be functionally independent and their level surfaces S_1 and S_2 intersect in a curve C. Since C then lies in both of the surfaces, S_1 and S_2, the tangent to C is normal to both $\vec{n}_1$ and $\vec{n}_2$, that is to both $\nabla\phi_1$ and $\nabla\phi_2$. This is the same thing as saying both ϕ_1 and ϕ_2 satisfy. We will illustrate with examples.

Example:

- Consider the radial vector field $\vec{V} = (t,x,u)$. A first integral must satisfy:

$$t\partial_t\phi(t,x,u) + x\partial_x\phi(t,x,u) + u\partial_u\phi(t,x,u) = 0.$$

To obtain a solution, we consider the following system of ode's:

$$\frac{dt}{t} = \frac{dx}{x} = \frac{du}{u} \quad or \quad \frac{dt}{t} = \frac{dx}{x} \quad and \quad \frac{dx}{x} = \frac{du}{u}$$

then $\quad \dfrac{dt}{t} = \dfrac{dx}{x} \quad$ leads to $\quad \dfrac{x}{t} = C_1,$

and $\quad \dfrac{dx}{x} = \dfrac{du}{u} \quad$ implies $\quad \dfrac{x}{u} = C_2.$

That is, $\phi_1(t,x,u) = \dfrac{x}{t}$ and $\phi_2(t,x,u) = \dfrac{x}{u}$ are a pair of first integrals for $\vec{V} = (t,x,u)$. We can show that for any smooth function F of two variables, $\phi_3(t,x,u) = F(\phi_1(t,x,u),\phi_2(t,x,u))$ is also a first integral for $\vec{V}$ and ϕ_3 is then viewed as an implicit representation for the most general solution of the first integral pde.

If $\phi_1(t,x,u)$ and $\phi_2(t,x,u)$ are a pair of first integrals for $\vec{V}$ then $\phi_3(t,x,u) = F(\phi_1(t,x,u),\phi_2(t,x,u))$ where F denotes an arbitrary smooth function of two variables, is also a first integral for $\vec{V}$.

$\phi_1(t,x,u) = \dfrac{x}{t}$ and $\phi_2(t,x,u) = \dfrac{x}{u}$ are functionally independent.

- Consider the vector field $\vec{V} = (t,x,u)$. The trajectories are solution curves for the following system of ode's:

$$\frac{dt}{t} = \frac{dx}{u} = \frac{du}{x} \quad or \quad \frac{dt}{t} = \frac{du}{x} \quad and \quad \frac{dx}{u} = \frac{du}{x}$$

From $\dfrac{dx}{u} = \dfrac{du}{x}$ it follows that $x^2 = u^2 + C_1$. Then,

$$\frac{dt}{t} = \frac{du}{x} = \frac{du}{\sqrt{u^2 + C_1}} \quad and \quad t = C_2\left(u + \sqrt{u^2 + C_1}\right) = C_2(u+x).$$

Evidently, $\phi_1 = x^2 - u^2$ and $\phi_2 = \dfrac{t}{u+x}$ are a pair of first integrals for the vector field $\vec{V} = (t,x,u)$.

Each of these functions satisfies:

$$t\partial_t\phi(t,x,u) + u\partial_x\phi(t,x,u) + x\partial_u\phi(t,x,u) = 0,$$

and the most general solution of this equation can be written implicitly as $F\left(x^2 - u^2, \dfrac{t}{u+x}\right) = 0$, where F denotes an arbitrary smooth function of two variables.

Characteristics for Quasilinear PDE's of Order 1

We are aware now that C is a characteristic curve for the quasilinear pde (a quasilinear equation) if C is a trajectory for the vector field $\vec{V} = (a,b,-f)$. Then solutions for the pde can be obtained from first integrals for the vector field. However, we are not usually interested in finding the most general solution for the pde but are instead interested in finding certain particular solutions. For example, we shall be interested in finding a solution for a quasilinear equation that satisfies the additional condition that:

$$u(x,t) = g(x,t) \quad on \quad C_I : \begin{cases} x = x(\tau), \\ t = t(\tau). \end{cases}$$

where the curve C_I and the function g are given. A condition of this form is called a Cauchy condition, and the problem of finding a solution for a quasilinear equation that satisfies the Cauchy condition is called a Cauchy initial value problem.

Example:

1. Consider the following Cauchy problem:

$$\partial_t u(x,t) + 4\partial_x u(x,t) = 0, \quad u(x,0) = \frac{1}{1+x^2}.$$

Let C_0 denote a solution curve for, $\dfrac{dt}{ds} = 1$, $\dfrac{dx}{ds} = 4$. This pair of equations is equivalent to the single equation, $\dfrac{dx}{dt} = 4$, for which the solution is, $x(t) = 4t + x_0$. The solution curves C_0, are a family of straight lines all having the same slope. Then along any solution curve,

$$\partial_t u(x,t) + 4\partial_x u(x,t) = \partial_t u(x,t) + \partial_x u(x,t)\frac{dx}{dt} = \frac{du}{dt} = 0,$$

from which it follows that u is a constant on each such curve. A general solution for the partial differential equation is given by $u(x,t) = f(x - 4t)$ where $f = f(z)$ denotes an arbitrary smooth function of one variable. Then $u(x,0) = f(x)$ and this, combined with the Cauchy initial condition, leads to the solution:

$$u(x,t) = \frac{1}{1+(x-4t)^2}$$

for the Cauchy problem. The initial value $u_0 = u(x_0,0)$ of the solution at the point x_0 propagates along the line $x - 4t = x_0$; i.e., $u(x,t) = u_0$ at all points (x,t) such that $x - 4t = x_0$. As a result, if the initial data is specified only on the interval, say $0 < x < 10$, then the solution is determined only in the strip, $\sum = \{(x,t): 4t < x < 4t + 10, t > 0\}$. The strip Σ is the domain of influence of the initial interval $I = \{0 < x < 10\}$. The pde in this example is linear which leads to the result that the characteristic system of ode's uncouples. That is, we can solve the equation $x'(t) = 4$ separately from the equation $u'(t) = 0$.

2. Now consider a Cauchy problem for the variable coefficient equation:

$$\partial_t u(x,t) + xt\partial_x u(x,t) = 0, \quad u(x,0) = \sin(x).$$

The coefficients in this equation are functions of the independent variables in the problem but do not depend on the unknown function u. Hence the equation is a linear partial differential equation as was the equation in the previous example. The solution curves for the characteristic ode, $\dfrac{dx}{dt} = xt$ are given by,

$$\ln x = t^2/2 + c_0, \quad or \quad x = c_1 e^{t^2/2}.$$

Evidently, the solution curves are the level curves of $\phi(x,t) = xe^{-t^2/2}$ and since the pde reduces to the ode $u'(s) = 0$ along level curves of ϕ, the solution u of the partial differential equation is constant along these curves. The most general such solution has the form $u(x,t) = f(xe^{-t^2/2})$ for an arbitrary smooth function of one variable f. Using this in the initial condition leads to,

$$u(x,0) = f(x) = \sin(x)$$

and

$$u(x,t) = \sin(xe^{-t^2/2}).$$

Points worth noting about these two examples:

- The solution curves C_0 constructed for these two examples are not characteristic curves for the pde's in the examples, since they are plane curves lying in the x-t plane. In fact, they are the projections into the x-t plane of the characteristic curves for the pde's. Since the coefficients in the pde's in these linear examples do not depend on the solution u, the characteristic system can be uncoupled and we can solve for $x = x(s)$, $t = t(s)$ without solving for u. The solution curves are curves in the x-t plane. In order to distinguish these plane curves from the previously discussed characteristic curves in 3-space, we will refer to the plane solution curves as base characteristics and use the term space characteristics when referring to characteristic curves in 3-space.

- The domain of influence for an interval $I = [a,b]$ is the strip $\sum = \{(x,t) : t > 0, ae^{t^2/2} < x < be^{t^2/2}\}$ bounded by the base characteristics that originate at the endpoints of the initial interval.

- The space characteristic curves in these two examples are curves of the form $\{x = x(t), u = const\}$ i.e., they are plane curves lying in planes parallel to the x-t plane. This is due to the fact that in both examples, the partial differential equation is homogeneous and in such cases, the pde reduces to $du/dt = 0$ along base characteristics with the obvious result that solutions are constant along base characteristics and along space characteristics.

- Constant coefficient equations have straight line base characteristics while linear equations with variable coefficients have base characteristics which are curves. However even when the base characteristics are curved, the family of curves is coherent; i.e., base characteristics which originate at distinct points can never cross.

To see why characteristics for linear equations are coherent, let C_1 and C_2 denote distinct base characteristics of the equation:

$$a(t,x)\partial_t u(x,t) + b(t,x)\partial_x u(t,x,u) = f(t,x,u),$$

which cross at some point (x_0, t_0). Intersection at a point requires that the curves have non-coincident tangents at the point of intersection. But this is inconsistent with equation $a(t,x,u)\partial_t u(x,t) + b(t,x,u)\partial_x u(t,x) = f(t,x,u)$, which implies that:

$$\left(\frac{dt}{dx}\right)_{C_1} = \frac{a(x_0,t_0)}{b(x_0,t_0)} = \left(\frac{dt}{dx}\right)_{C_2}$$

i.e., the tangents to C_1 and C_2 have equal slopes at the point of intersection. In the case of quasilinear equations, where the coefficients can depend on u, we will see that this coherence can fail.

3. Consider a Cauchy problem for a linear but inhomogeneous equation:

$$x\partial_t u(x,t) - 2xt\partial_x u(x,t) = 2tu, \quad u(x,0) = x .$$

The characteristics are the solutions of:

$$\frac{dt}{ds} = x, \; \frac{dx}{ds} = -2xt, \; \frac{du}{ds} = 2tu, \; or \; \frac{dt}{x} = -\frac{dx}{2xt}, \; and \; \frac{du}{2tu} = -\frac{sx}{2xt}.$$

Since the equation is linear, the base characteristic curves can be obtained independent of the solution u. The solutions for the characteristic equations are given by:

$$x + t^2 = C_0 \quad and \quad xu = C_1.$$

The general solution for the partial differential equation can be expressed as:

$$u = \frac{C_1}{x} = \frac{F(x+t^2)}{x}$$

and then $u(x,0) = \dfrac{F(x)}{x} = x^3$. It follows that $F(x) = x^4$ hence:

$$u(x,t) = \frac{(x+t^2)^4}{x}.$$

When the partial differential equation is inhomogeneous, as in this example, the solution u is no longer constant along the characteristics, but in fact varies in a manner prescribed by the differential equation, $u'(s) = 2tu$, and the initial condition. Then the space characteristics are no longer plane curves but are truly space curves. The partial differential equation reduces to an ordinary differential equation along the characteristics.

4. Now consider the following quasilinear problem,

$$\partial_t u(x,t) + 4\partial_x u(x,t) = u(x,t)^2, \quad u(x,0) = \frac{1}{1+x^2}.$$

The characteristic system can be written as:

$$\frac{dt}{1} = \frac{dx}{4} = \frac{du}{u^2} = \; or \; \frac{dt}{1} = \frac{dx}{4} \; and \; \frac{dt}{1} = \frac{du}{u^2}.$$

The pde is linear in the leading terms (in fact, this special subclass of quasilinear problems is referred to as the class of semilinear problems) so that the characteristic system uncouples and the base characteristics can be found without knowing the solution u. Clearly the base characteristics are the family of straight lines $x - 4t = x_0$. The second characteristic equation $u'(t) = u(t)^2$

has the solution $u(t) = (C_1 - t)^{-1}$ and it follows that the most general solution for the pde can be written as:

$$u(x,t)\frac{1}{f(x-4t)-t}.$$

The initial condition implies $u(x,0) = \dfrac{1}{f(x)} = \dfrac{1}{1+x^2}$, or $f(x) = 1 + x^2$. Then the solution of the Cauchy initial value problem is:

$$u(x,t) = \frac{1}{(x-4t)^2 + 1 - t}$$

In spite of the smoothness of the initial data, the solution develops a singularity at $x = 4, t = 1$. Plotting solution profiles at times approaching $t = 1$ shows that the solution behaves like a wave that propagates from left to right and sharpens to a spike at $x = 4$ as t approaches $t = 1$.

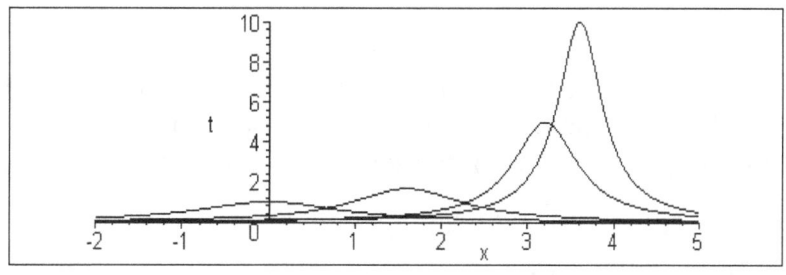

u(x,t) vs x for t=0, .2, .6, .9.

We compute:

$$\partial_t u(x,t) = \frac{-2(x-4t)}{(1-t+(x-4t)^2)^2}$$

from which it is evident that the gradient, $\partial_x u(x,t)$ has an even stronger singularity at $x = 4, t = 1$. The developing gradient singularity can be seen in the following plot of gradient profiles for increasing t.

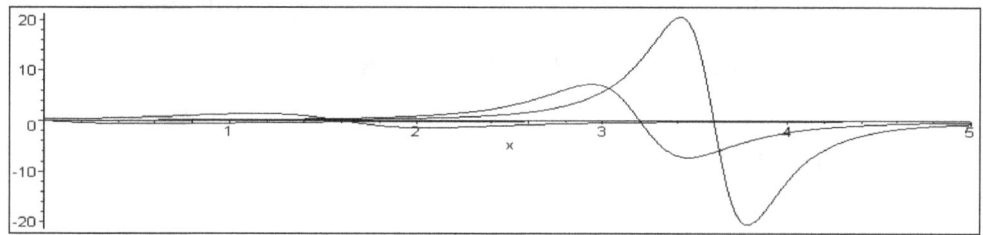

Gradient of u(x,t) vs x for t=0, .2, .6, .9.

Since this pde is semilinear (i.e., linear in its leading terms) and has smooth initial data, the spontaneous singular behavior in the solution must be due to the nonlinear term on the right side of the equation. Note also that semilinear equations permit the base characteristics (i.e. the solution curves $t'(s) = a$, $x'(s) = b$ to be found independent of the solution $u = u(x,t)$ and then the PDE reduces to a nonlinear ODE along the base characteristics.

5. Consider the quasilinear Cauchy problem:

$$\partial_t u(x,t) u \partial_x u(x,t) = 0, \quad u(x,0) = \frac{1}{1+x^2} = f(x).$$

Here the coefficients of $\partial_t u(x,t) \ and \ \partial_x u(x,t)$ depend on u so the characteristic equations become:

$$\frac{dt}{ds} = 1, \ \frac{dx}{ds} = u \ and \ \frac{du}{ds} = 0 \ or \ \frac{dx}{dt} = u \ and \ \frac{du}{dt} = 0.$$

In this case these equations do not uncouple. In order to solve the equations as a system, we note first the equation $u'(t) = 0$. This implies $u = const$ along solutions of the other equation, $dx/dt = u$. This is to say, $u = u_0$ along the straight lines, $x - u_0 t = x_0$. It follows from the initial condition that the initial value u_0 originating at $(x_0,0)$ is given by $u(x_0,0) = u_0 = f(x_0)$. The solution u can then be expressed implicitly by writing:

$$u(x,t) = \frac{1}{1+(x-ut)^2}$$

and this equation can be solved for u in terms of x and t. The result, obtainable using Maple or Mathematica, is a complicated function of (x,t). As in the previous example, the quasilinearity of the pde produces some singular behavior in the solution. For instance, suppose:

$$u = u_0 = f(x_0) \ along \ x - u_0 t = x_0$$

and $\qquad u = u_1 = f(x_1) \ along \ x - u_1 t = x_1 \neq x_0.$

If $u_1 \neq u_0$ then the two straight lines intersect at (x_*, t_*) where

$$t_* = -\frac{x_1 - x_0}{u_1 - u_0}, \quad x_* = \frac{u_1 x_0 - u_0 x_1}{u_1 - u_0}.$$

The time of intersection, t_*, is positive if $x_1 > x_0$ and $u_1 < u_0$. At such a point of intersection, $u(x_*, t_*)$ has the impossible requirement of being simultaneously equal to the distinct values $u_1 \neq u_0$. We conclude that the solution breaks down in some way at this point. Note further that:

$$u(x,t) = f(x-ut) \ leads \ to \ \partial_x u(x,t) = f'(x-ut)(1-t\partial_x u(x,t));$$

i.e.,

$$\partial_x u(x,t) = \frac{f'(x-ut)}{1-tf'(x-ut)}.$$

Evidently the gradient $\partial_x u(x,t)$ becomes undefined at any point where, $1-tf'(x-ut)=0$, another indication that the solution breaks down at some finite time.

By plotting the solution profiles versus x for a sequence of increasing times, it can be seen that the initial wave form moves to the right, deforming as it propagates. It is apparent that at some point, the tangent line to the profile becomes vertical and the solution is no longer single valued beyond this point.

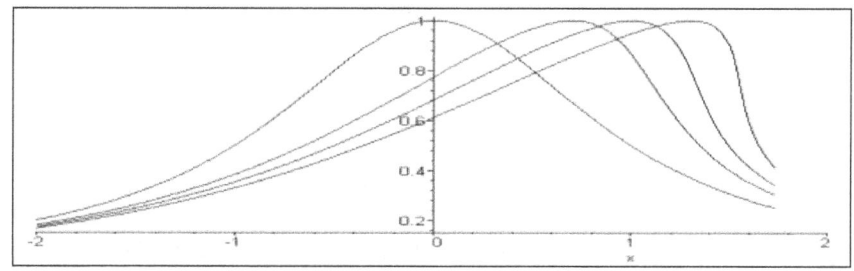

Solution Profiles at t=0, 1/2, 1, 3/2.

The spontaneously occuring singularity is also evident from plotting the gradient versus x at increasing times. We can see that as time increases, the gradient begins to develop a negatively infinite singularity as the tangent line tends toward the vertical. Here, as in the previous example, the singularity is due to the nonlinearity of the equation (which in this case leads to colliding characteristics).

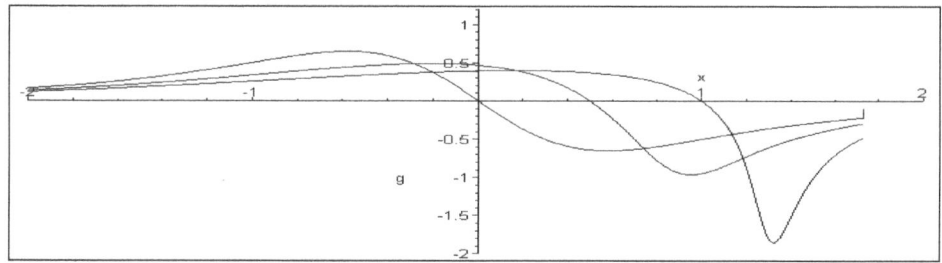

Gradient Profiles at t=0, 1/2, 1, 3/2.

In each of the last two examples we have seen equations with smooth coefficients and initial data develop spontaneous singularities due to the nonlinearity of the equations. The solutions in these two examples break down at some finite time and no classical solution for the initial value problems exists past this point of breakdown. It will be necessary to weaken the notion of solution in order for these nonlinear problems to be solvable globally.

Wave Equation with Constant Speed

Consider the first-order wave equation with constant speed:

$$\frac{\partial u}{\partial t} + c\frac{\partial u}{\partial x} = 0.$$

It responds well to a change of variables:

$$\xi = x + ct \qquad \eta = x - ct$$

The chain rule gives us:

$$\frac{\partial}{\partial x} = \frac{\partial \xi}{\partial x}\frac{\partial}{\partial \xi} + \frac{\partial \eta}{\partial x}\frac{\partial}{\partial \eta} = \frac{\partial}{\partial \xi} + \frac{\partial}{\partial \eta}$$

$$\frac{\partial}{\partial t} = \frac{\partial \xi}{\partial t}\frac{\partial}{\partial \xi} + \frac{\partial \eta}{\partial t}\frac{\partial}{\partial \eta} = c\left(\frac{\partial}{\partial \xi} - \frac{\partial}{\partial \eta}\right)$$

and so the wave equation is equivalent to:

$$2c\frac{\partial u}{\partial \xi}=0.$$

Integrating gives the general solution $u=F(\eta)$, $u=F(x-ct)$.

But where did we get the change of variables from? The line $x-ct=$ constant is a line in the $x-t$ plane along which u is constant. This means that if we parametrise this line:

$$x=x(r) \quad t=t(r)$$

then moving along the line by changing r will not change u, i.e.

$$\frac{du}{dr}=0.$$

This is the underlying principle of the characteristic.

Variable Speed

Let's look now at the variable speed case:

$$\frac{\partial u}{\partial t}+c(x,t)\frac{\partial u}{\partial x}=0.$$

We would like again to find curves along which u is constant. Suppose such a curve is given by $x=x(r)$ and $t=t(r)$. Then, using the chain rule,

$$\frac{du}{dr}=\frac{\partial u}{\partial t}\frac{dt}{dr}+\frac{\partial u}{\partial x}\frac{dx}{dr}.$$

We want this to be zero, which is easily achieved if we make this expression the same as the original linear operator:

$$\frac{dt}{dr}\frac{\partial u}{\partial t}+\frac{dx}{dr}\frac{\partial u}{\partial x}=\frac{\partial u}{\partial t}+c(x,\,t)\frac{\partial u}{\partial x}=0.$$

This gives us the two parametric equations governing the shape of the characteristic curve:

$$\frac{dt}{dr}=1, \quad \frac{dx}{dr}=c(x,r).$$

These are both ODEs and straightforward to solve.

Example: Look at the equation,

$$2\sin\theta\,\cos 2\phi\frac{\partial u}{\partial \theta}-\frac{\cos\theta\sin 2\phi}{\sin\theta}\frac{\partial u}{\partial \phi}=0.$$

Suppose that our characteristic is given by $\theta = \theta(r), \phi = \phi(r)$. Then the requirement that u be constant along a characteristic becomes:

$$\frac{\partial u}{\partial \theta}\frac{d\theta}{dr} + \frac{\partial u}{\partial \phi}\frac{d\phi}{dr} = 0.$$

A naïve attempt would be to look at the coupled ODEs:

$$\frac{d\theta}{dr} = 2\sin\theta\cos 2\phi \qquad \frac{d\phi}{dr} = -\frac{\cos\theta\sin 2\phi}{\sin\theta}$$

but we can uncouple them if, before we start, we multiply the original equation by $\sin\theta/\cos\theta\cos 2\phi$:

$$\frac{2\sin^2\theta}{\cos\theta}\frac{\partial u}{\partial \theta} - \frac{\sin 2\phi}{\cos 2\phi}\frac{\partial u}{\partial \phi} = 0,$$

$$\frac{d\theta}{dr} = \frac{2\sin^2\theta}{\cos\theta} \qquad \frac{d\phi}{dr} = -\frac{\sin 2\phi}{\cos 2\phi}.$$

Now the equations are decoupled, and solving them in turn gives:

$$\sin\theta = -\frac{1}{2r} \qquad \sin 2\phi = \exp[C - 2r].$$

We only use a constant of integration in one of these equations; since r is just a parameter, the point r = 0 is not defined a priori. Effectively, we are making a change of variables from x, t to r, C. We can invert the transformation:

$$C = \ln\sin 2\phi - \frac{1}{\sin\theta} \qquad r = -\frac{1}{2\sin\theta}$$

and since u is constant on this curve, we can deduce the general solution:

$$u = F(C) = F\left(\ln\sin 2\phi - \frac{1}{\sin\theta}\right).$$

More than Two Dimensions

Now suppose we have the PDE:

$$\frac{\partial u}{\partial x} + c_1(x,\, y,\, z)\frac{\partial u}{\partial y} + c_2(x,\, y,\, z)\frac{\partial u}{\partial z} = 0.$$

Again, we look for a curve on which u is constant; being a curve, it can still be described with a single variable $x = x(r),\ y = y(r)$ and $z = z(r)$. Then the chain rule gives:

$$\frac{du}{dr} = \frac{\partial u}{\partial x}\frac{dx}{dr} + \frac{\partial u}{\partial y}\frac{dy}{dr} + \frac{\partial u}{\partial z}\frac{dz}{dr}$$

and to make this equal to zero we choose:

$$\frac{dx}{dr}=1 \qquad \frac{dy}{dr}=c_1(x(r),\,y(r),\,z(r)) \qquad \frac{dz}{dr}=c_2(x(r),\,y(r),\,z(r)).$$

The latter two are now coupled ODEs so we are not guaranteed to be able to find a solution; but sometimes you may be lucky.

Example: Look at the equation,

$$\frac{\partial u}{\partial x}+xy\frac{\partial u}{\partial y}+2x^2z\ln y\frac{\partial u}{\partial z}=0.$$

We set $x=x(r)$, $y=y(r)$ and $z=z(r)$ and the chain rule gives:

$$\frac{du}{dr}=\frac{\partial u}{\partial x}\frac{dx}{dr}+\frac{\partial u}{\partial y}\frac{dy}{dr}+\frac{\partial u}{\partial z}\frac{dz}{dr}.$$

To match the three coefficients we set:

$$\frac{dx}{dr}=1 \qquad x(r)=r$$

$$\frac{dy}{dr}=xy=ry \qquad y(r)=y_0\,\exp\left[r^2/2\right]$$

$$\frac{dz}{dr}=2x^2z\ln y=-r^4z\ln y_0 \qquad z(r)=z_0\,\exp\left[-r^5\ln y_0/5\right].$$

Now we have expressed all points in terms of the three parameters r,y_0 and z_0 and u is independent of r, so the solution is any function of y_0 and z_0. Reversing the change of variables gives:

$$r=x \qquad y_0=y\,\exp\left[-x^2/2\right] \qquad z_0=z\,\exp\left[-x^7/10\right]y^{\left[x^5/5\right]}$$

and the full solution is:

$$u=F\left(y\,\exp\left[-x^2/2\right];\,z\,\exp\left[-x^7/10\right]y^{\left[x^5/5\right]}\right).$$

First Order Linear Partial Differential Equations

The most general first-order linear PDE has the form:

$$a(x,y)z_x+b(x,y)z_y+c(x,y)z_z=d(x,y),$$

where a, b, c, and d are given functions of x and y. These functions are assumed to be continuously differentiable. Rewriting above equation as:

$$a(x,y)z_x+b(x,y)z_y=-c(x,y)z+d(x,y),$$

we observe that the left hand side of the above i.e.,

$$a(x, y)z_x + b(x, y)z_y = \nabla z \cdot (a, b)$$

is (essentially) a directional derivative of z(x, y) in the direction of the vector (a, b), where (a, b) is defined and nonzero. When a and b are constants, the vector (a, b) had a fixed direction and magnitude, but now the vector can change as its base point (x, y) varies. Thus, (a, b) is a vector field on the plane.

The equations:

$$\frac{dx}{dt} = a(x,y), \quad \frac{dy}{dt} = b(x,y)$$

determine a family of curves x = x(t), y = y(t) whose tangent vector $\left(\dfrac{dx}{dt}, \dfrac{dy}{dt}\right)$ coincides with the direction of the vector (a, b). Therefore, the derivative of z(x, y) along these curves becomes:

$$\frac{dz}{dt} = \frac{d}{dt}z\{(x(t),y(t))\} = \frac{\partial z}{\partial x}\frac{dx}{dt} + \frac{\partial z}{\partial y}\frac{dy}{dt}$$

$$= z_x\big(x(t), y(t)\big)a\big(x(t), y(t)\big) + z_y\big(x(t), y(t)\big)b\big(x(t), y(t)\big)$$

$$= -c\big(x(t), y(t)\big)z\big(x(t), y(t)\big) + d\big(x(t), y(t)\big)$$

$$= -c(t)z(t) + d(t),$$

where we have used the chain rule and equation $a(x, y)z_x + b(x, y)z_y + c(x, y)z = d(x, y)$. Thus, along these curves, z(t) = z(x(t), y(t)) satisfies the ODE:

$$z'(t) + c(t)z(t) = d(t).$$

Let $\mu(t) = \exp\left[\int_0^t c(\tau)d\tau\right]$ be an integrating factor for equation $z'(t) + c(t)z(t) = d(t)$.

Then, the solution is given by:

$$z(t) = \frac{1}{\mu(t)}\left[\int_0^t \mu(\tau)d(\tau)d\tau + z(0)\right].$$

The approach described above to solve equation $a(x, y)z_x + b(x, y)z_y + c(x, y)z = d(x, y)$, by using the solutions of equations:

$$\frac{dx}{dt} = a(x,y), \quad \frac{dy}{dt} = b(x,y)$$

and

$$z'(t) + c(t)z(t) = d(t),$$

is called the method of characteristics. It is based on the geometric interpretation of the partial differential equation $a(x, y)z_x + b(x, y)z_y + c(x, y)z = d(x, y)$.

(i) The ODEs is known as the characteristics equation for the PDE. The solution curves of the characteristic equation are the characteristics curves for equation:

$$a(x, y)z_x + b(x, y)z_y + c(x, y)_z = d(x, y).$$

(ii) Observe that $\mu(t)$ and $d(t)$ depend only on the values of $c(x, y)$ and $d(x, y)$ along the characteristics curve $x = x(t), y = y(t)$. Thus, equation $z(t) = \dfrac{1}{\mu(t)}\left[\displaystyle\int_0^t \mu(\tau)d(\tau)d\tau + z(0)\right]$ shows that the values $z(t)$ of the solution z along the entire characteristics curve are completely determined, once the value $z(0) = z(x(0), y(0))$ is prescribed.

(iii) Assuming certain smoothness conditions on the functions a, b, c, and d, the existence and uniqueness theory for ODEs guarantees a unique solution curve $(x(t), y(t), z(t))$ of equations:

$$\frac{dx}{dt} = a(x,y), \ \frac{dy}{dt} = b(x,y)$$

and,

$$z'(t) + c(t)z(t) = d(t), \quad \text{(i.e., a characteristic curve) passes through a given point}$$

(x_0, y_0, z_0) in (x, y, z) space.

The Method of Characteristics for Solving Linear First-order IVP

In practice we are not interested in determining a general solution of the partial differential equation $a(x, y)z_x + b(x, y)z_y + c(x, y)_z = d(x, y)$, but rather a specific solution z = z(x, y) that passes through or contains a given curve C. This problem is known as the initial value problem for the above equation. The method of characteristics for solving the initial value problem for equation $a(x, y)z_x + b(x, y)z_y + c(x, y)_z = d(x, y)$, proceeds as follows.

Let the initial curve C be given parametrically as:

$$x = x(s), y = y(s), z = z(s).$$

for a given range of values of the parameter s. The curve may be of finite or infinite extent and is required to have a continuous tangent vector at each point.

Every value of s fixes a point on C through which a unique characteristic curve passes. The family of characteristic curves determined by the points of C may be parameterized as:

$$x = x(s, t), y = y(s, t), z = z(s, t)$$

with t = o corresponding to the initial curve C. That is, we have:

$$x(s, 0) = x(s), y(s, 0) = y(s), z(s, 0) = z(s).$$

In other words, we have the following:

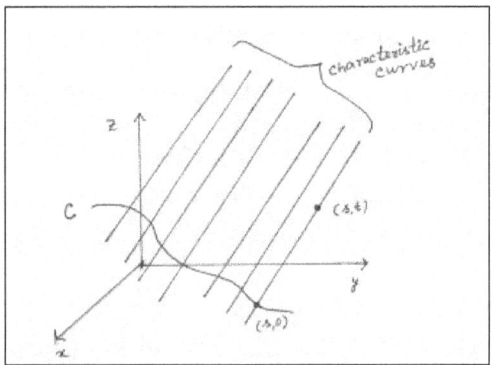

Characteristic curves and construction of the integral surface.

The functions x(s, t) and y(s, t) are the solutions of the characteristics system (for each fixed s):

$$\frac{d}{dt}x(s,t) = a\big(x(s,t),y(s,t)\big), \frac{d}{dt}y(s,t) = b\big(x(s,t),y(s,t)\big)$$

with given initial values x(s, 0) and y(s, 0).

Suppose that,

$$z\big(x(s,\ 0),\ y(s,\ 0)\big)\ =\ g(s),$$

where g(s) is a given function. We obtain $z\big(x(s,\ t),\ y(s,\ t)\big)$ as follows: Let,

$$z(s,t) = z\big(x(s,t),y(s,t)\big),\ c(s,t) = c\big(x(s,t),y(s,t)\big),\ d(s,t) = d\big(x(s,t),\ y(s,t)\big)$$

and

$$\mu(s,t) = \exp\left[\int_0^t c(s,t)dt\right].$$

Analogous to formula $z(t)\ =\ \dfrac{1}{\mu(t)}\left[\int_0^t \mu(\tau)d(\tau)d\tau\ +\ z(0)\right]$, for each fixed s, we obtain:

$$z(s,t) = \frac{1}{\mu(s,t)}\left[\int_0^t \mu(s,t)d(s,t)dt + g(s)\right].$$

$z(s,\ t)$ is the value of z at the point $x(s,\ t),\ y(s,\ t))$. Thus, as s and t vary, the point $(x,\ y,\ z)$, in xyz-space, given by:

$$x\ =\ x(s,\ t),\ y\ =\ y(s,\ t),\ z\ =\ z(s,\ t),$$

traces out the surface of the graph of the solution z of the PDE $a(x,\ y)z_x\ +\ b(x,\ y)z_y\ +\ c(x,\ y)_z$ $=\ d(x,\ y)$, which meets the initial curve ($z\big(x(s,\ 0),\ y(s,\ 0)\big)\ =\ g(s)$). The equation $x\ =\ x(s,\ t)$,

$y = y(s, t)$, $z = z(s, t)$ constitute the parametric form of the solution of equation $a(x, y)z_x$ $+ b(x, y)z_y + c(x, y)_z = d(x, y)$, satisfying the initial condition of equation $z(x(s, 0), y(s, 0))$ $= g(s)$, [i.e., a surface in (x, y, z)-space that contains the initial curve].

If the Jacobian $J(s, t) = x_s y_t - x_t y_s \neq 0$, then the equations $x = x(s, t)$ and $y = y(s, t)$ can be inverted to give s and t as (smooth) functions of x and y i.e., s = s(x, y) and t = t(x, y). The resulting function $z = z(x, y) = z(s(x, y), t(x, y))$ satisfies the PDE (1) in a neighborhood of the curve C (in view of equation $z'(t) + c(t)z(t) = d(t)$ and the initial condition of equation $x = x(s)$, $y = y(s)$, $z = z(s)$.) and is the unique solution of the IVP.

Example: Determine the solution the following IVP:

$$\frac{\partial z}{\partial y} + c\frac{\partial z}{\partial x} = 0, \; z(x, 0) = f(x),$$

where f(x) is a given function and c is a constant.

Solution: A step by step procedure for the finding solution is given below.

Step 1: (Finding characteristic curves)

To apply the method of characteristics, parameterize the initial curve C as follows: as follows:

$$x = s, y = 0, z = f(s).$$

The family of characteristics curves x((s, t), y(s, t)) are determined by solving the ODEs:

$$\frac{d}{dt}x(s, t) = c, \quad \frac{d}{dt}y(s, t) = 1$$

The solution of the system is:

$$x(s,t) = ct + c_1(s) \text{ and } y(s, t) = t + c_2(s).$$

Step 2: (Applying IC)

Using the initial conditions:

$$x(s,0) = s, \quad y(s,0) = 0.$$

we find that:

$$c_1(s) = s, c_2(s) = 0,$$

and hence:

$$x(s,t) = ct + s \text{ and } y(s,t) = t.$$

Step 3: (Writing the parametric form of the solution)

Comparing with equation $a(x, y)z_x + b(x, y)z_y + c(x, y)_z = d(x, y)$, we have $c(x, y) = 0$ and

$d(x, y) = 0$. Therefore, using equation $\mu(s,t) = \exp\left[\int_0^t c(s,t)dt\right]$ and

$$z(s,t) = \frac{1}{\mu(s,t)}\left[\int_0^t \mu(s,t)d(s,t)dt + g(s)\right],$$

we find that:

$$d(s, t) = 0, \ \mu(s, t) = 1.$$

Step 4 : (Expressing z(s, t) in terms of z(x, y)) Expressing s and t as $s = s(x, y)$ and $t = t(x, y)$,

$$s = x - cy, \ t = y.$$

We now write the solution in the explicit form as:

$$z(x, y) = z(s(x, y), \ y(x, y)) = f(x - cy).$$

Clearly, if f(x) is differentiable, the solution $z(x, y) = f(x - cy)$ satisfies given PDE as well as the initial condition.

Example: Characterizes unidirectional wave motion with velocity c. If we consider the initial function $z(x, 0) = f(x)$ to represent a waveform, the solution $z(x, y) = f(x - cy)$ shows that a point x for which $x - cy = $ constant, will always occupy the same position on the wave form. If c > 0, the entire initial wave form $f(x)$ moves to the right without changing its shape with speed c (if c < 0, the direction of motion is reversed).

Example: Find the parametric form of the solution of the problem.

$$-yz_x + xz_y = 0$$

with the condition given by:

$$z(s, s^2) = s^3, \ (s > 0).$$

Solution: To find the solution, let's proceed as follows.

Step 5: (Finding characteristic curves)

The family of characteristics curves $(x(s, t), y(s, t))$ are determined by solving:

$$\frac{d}{dt}x(s, t) = -y(s, t), \ \frac{d}{dt}y(s, t) = x(s, t)$$

with initial conditions:

$$x(s, 0) = s, \ y(s, 0) = s^2.$$

The general solution of the system is:

$$x(s, t) = c_1(s) \cos(t) + c_2(s) \sin(t) \text{ and } y(s, t) = c_1(s)\sin(t) - c_2(s) \cos(t).$$

Step 6: (Applying IC)

Using ICs, we find that:

$$c_1(s) = s, \quad c_2(s) = -s^2,$$

and hence:

$$x(s,t) = s \cos(t) - s^2 \sin(t) \text{ and } y(s,t) = s \sin(t) + s^2 \cos(t).$$

Step 7: (Writing the parametric form of the solution)

Comparing with equation $a(x, y)z_x + b(x, y)z_y + c(x, y)_z = d(x, y)$, we note that $c(x, y) = 0$ and $d(x, y) = 0.$ Therefore, using equation $\mu(s,t) = \exp\left[\int_0^t c(s,t)dt\right]$ and $z(s,t) = \dfrac{1}{\mu(s,t)}\left[\int_0^t \mu(s,t)d(s,t)dt + g(s)\right]$, it follows that

$$d(s, t) = 0, \mu(s, t) = 1.$$

In view of the given condition curve and $z = z(s, t)$, we obtain

$$z(x(s, 0), y(s, 0)) = z(s, s^2) = g(s) = s^3, z(s, t) = s^3.$$

Thus, the parametric form of the solution of the problem is given by

$$x(s, t) = s \cos(t) - s^2 \sin(t), y(s, t) = s \sin(t) + s^2 \cos(t), z(s, t) = s^3.$$

Step 8: (Expressing z(s, t) in terms of z(x, y))

Writing s and t as a function of x and y, it is an easy exercise to show that

$$z(x, y) = \frac{1}{\sqrt{8}}\left[-1 + \sqrt{1 + 4(x^2 + y^2)}\right]^{3/2}.$$

First Order Non-linear Partial Differential Equations

Nonlinear First-Order PDE

Let $x = (x_1, ..., x_n) \in \mathbb{R}^n$ and let $u : \overline{\Omega} \subseteq \mathbb{R}^n \to \mathbb{R}$. A first order partial differential equation for $u = u(x)$ is given by $F(Du, u, x) = 0$ where $F : \mathbb{R}^n \times \mathbb{R} \times \overline{\Omega} \to \mathbb{R}$ is a given function, and Du is the vector of partial derivatives of u.

We usually denote the entries of F as follows: $F = F(p, z, x)$. Thus $p \in \mathbb{R}^n$, $z \in \mathbb{R}$, $x \in \bar{\Omega}$.

The PDE $F(Du, u, x) = 0$ is usually accompanied by a boundary condition of the form $u = g$ on $\partial\Omega$. Such a problem is usually called a boundary value problem.

Example: (The eikonal equation). The eikonal equation,

$$|Du| = 1,$$

introduced by Hamilton in 1827 is an approximation to the equations which govern the behavior of light traveling through varying materials. A solution, depending on parameters $\|a\| = 1$, $b \in \mathbb{R}$ is:

$$u(x; a, b) = a \cdot x + b.$$

Example: (The Hamilton-Jacobi equation). The (simple version of the) Hamilton-Jacobi equation:

$$u_t + H(Du) = 0,$$

with $H : \mathbb{R}^n \to \mathbb{R}$ is an important equation from mechanics. $u = u(x, t) : \mathbb{R}^n \times \mathbb{R} \to \mathbb{R}$. A solution, depending on parameters $a \in \mathbb{R}^n$, $b \in \mathbb{R}$ is:

$$u(x, t; a, b) = a \cdot x - tH(a) + b$$

where $t \geq 0$.

For simplicity, in most of what follows, we restrict to $n = 2$. We call the two variables x, y. Thus, we reduce to the case:

$$F(u_x, u_y, u, x, y) = 0.$$

In this case, the solution $u = u(x, y)$ is a surface in $\mathbb{R}^3$. The normal direction to the surface at each point is given by the vector $(u_x, u_y, -1)$. We will use this fact later, when we construct our solutions using geometric methods.

Linear Equations

The first subclass of nonlinear first-order PDE we consider are the linear equations, which have the form:

$$a(x,y)u_x + b(x,y)u_y = c_0(x,y)u + c_1(x,y).$$

Example: Consider the linear equation:

$$u_x = c_0 u + c_1(x,y).$$

where c_0 is a constant, and $c_1(x,y)$ is a function of the two variables x and y. Since there is no differentiation with respect to y, the y variable may actually be treated as a parameter, and this equation reduces to an ODE.

To solve, we need some initial condition, for example:

$$u(0,y) = y.$$

The solution is now simple:

$$u(x,y) = e^{c_0 x}\left(\int_0^x e^{-c_0\xi}c_1(\xi,y)d\xi + y\right).$$

What do we learn from this? We see that given an initial value along the y-axis, $u(0, y_0)$, we are able to extend to a solution $u(\cdot, y_0)$ parallel to the x-axis. This is true for any y_0.

Does this always work? We split into cases.

- If for each y_0 we have one piece of information, say at $(0, y_0)$ as above, then we get a unique solution to the ODE along the line through y_0 that is parallel to the x-axis. Thus there is a unique solution to the PDE.

- If the initial data is given along a different curve (i.e. not the y-axis necessarily) which has the same y value at two different x values, but with values that do not permit a solution to the ODE along that y value, then there is no solution to the PDE as well.

- If we are given an "insufficient" amount of information (for example, if our initial data is given on a line parallel to the x-axis and precisely overlaps a solution to the ODE) then we have an infinite number of solutions to the PDE.

Quasi-Linear Equations

Another special case of nonlinear first-order PDE are the quasi-linear equations, where the non-linearity appears only on the z (that is, u) variable of F. The general form of such an equation is:

$$a(x, y, u)u_x + b(x, y, u)u_y = c(x, y, u).$$

The Method of Characteristics

The method of characteristics, developed by Hamilton in the 19th century, is essentially the method described above, only for more general examples: We want to construct the two-dimensional surface $u(x,y)$ in $\mathbb{R}^3$ that is a solution to our boundary value problem by translating the problem into infinitely many first order ODEs with initial data inherited from the data provided for the PDE.

Linear Equations

We return to the linear equation and write the initial data in parametric form:

$$\Gamma = \left(x_0(s), y_0(s), u_0(s)\right)$$

Where $s \in I = (\alpha, \beta)$.

Γ is called the initial curve.

Now, we rewrite equation $a(x,y)u_x + b(x,y)u_y = c_0(x,y)u + c_1(x,y)$ as:

$$\left(a,\ b,\ c_0 u + c_1\right) \cdot \left(u_x,\ u_y, -1\right) = 0.$$

As $\left(u_x,\ u_y, -1\right)$ is perpendicular to the solution surface, the vector $\left(a,\ b,\ c_0 u + c_1\right)$ must always lie in the tangent plane to the surface. Thus, the system of equations:

$$\dot{x}(t) = a\left(x(t), y(t)\right)$$
$$\dot{y}(t) = b\left(x(t), y(t)\right)$$
$$\dot{u}(t) = c_0\left(x(t), y(t)\right)u + c_1\left(x(t), y(t)\right),$$

defines a family of curves that lie in the solution surface. These equations are called the characteristic equations and the curves are called the characteristics of the PDE. This is an autonomous system, i.e. there's no dependence on the t variable (which makes sense, since there's no "dynamic" aspect to this problem).

Each characteristic must "start" at the initial curve, and thus we denote the characteristics by $\left(x(t, s), y(t, s), u(t, s)\right)$ where the variable s tracks the starting point $\Gamma(s)$ on the initial curve:

$$x(0, s) = x_0(s), y(0, s) = y_0(s), u(0, s) = u_0(s)$$

To summarize: The parameter t measures how "far" we are from the initial curve. For $t = 0$ we are actually on it. The parameter s measures where we are along the initial curve. The resulting surface is called an integral surface.

Indeed, from calculus we know that (generally) two parameters describe a surface in $\mathbb{R}^3$.

General Nonlinear Equations

Let us derive the characteristic equations in full generality. We once again consider:

$$F\left(Du, u, x\right) = 0$$

where $x \in \Omega \subseteq \mathbb{R}^n$, with the boundary condition:

$$u = g \text{ on } \Gamma$$

where $\Gamma \subseteq \partial\Omega$ and $g : \Gamma \to \mathbb{R}$ are given. We assume F, g to be smooth as well.

Suppose $x(t) = \left(x^1(t), \ldots, x^n(t)\right)$ is a parametric representation of a characteristic curve (which will be calculated later). We define two functions of the parameter t that track the values of u and Du along the characteristic:

$$z(t) := u\left(x(t)\right),$$

and

$$p(t) := Du(x(t)),$$

so that $p(t) = \left(p^1(t), \ldots, p^n(t)\right)$

where $p^i(t) = u_{x_i}(x(t)).$

Differentiating the above equation we have:

$$p^i(t) = \sum_{j=1}^{n} u_{x_i x_j}\left(x(t)\right) \dot{x}^j(t).$$

Differentiating equation $F(Du, u, x) = 0$ with respect to x_i, we have:

$$\sum_{j=1}^{n} \frac{\partial F}{\partial p_j}(Du, u, x)u_{x_j x_i} + \frac{\partial F}{\partial z}(Du, u, x)u_{x_i} + \frac{\partial F}{\partial x_i}(Du, u, x) = 0.$$

Evaluating above equation at $x = x(t)$, we get (using equations $z(t) := u(x(t))$, and $p(t) := Du(x(t))$, the identity:

$$\sum_{j=1}^{n} \frac{\partial F}{\partial p_j}(p(t), z(t), x(t))\, u_{x_i x_j}(x(t)) + \frac{\partial F}{\partial z}(p(t), z(t), x(t))\, p^i(t) + \frac{\partial F}{\partial x_i}(p(t), z(t), x(t)) = 0.$$

We also make the assumption:

$$\dot{x}^j(t) = \frac{\partial F}{\partial p_j}(p(t), z(t), x(t)),$$

for $j = 1, \ldots, n.$

Substituting equations:

$$\sum_{j=1}^{n} \frac{\partial F}{\partial p_j}(p(t), z(t), x(t))\, u_{x_i x_j}(x(t)) + \frac{\partial F}{\partial z}(p(t), z(t), x(t))\, p^i(t) + \frac{\partial F}{\partial x_i}(p(t), z(t), x(t)) = 0.$$

and

$$\dot{x}^j(t) = \frac{\partial F}{\partial p_j}(p(t), z(t), x(t)), \text{ into } p\ddot{y}i(t) = \sum_{j=1}^{n} u_{x_i x_j}(x(t))\, \dot{x}^j(t).$$

we have:

$$\dot{p}^i(t) = -\frac{\partial F}{\partial x_i}(p(t), z(t), x(t)) - \frac{\partial F}{\partial z}(p(t), z(t), x(t))\, p^i(t)$$

for i = 1, ..., n. Last, we differentiate equation $z(t) := u(x(t)),$ to get

$$\dot{z}(t) = \sum_{j=1}^{n} \frac{\partial u}{\partial x_j}\big(\mathrm{x}(t)\big)\, \dot{x}^j(t) = \sum_{j=1}^{n} p^j(t)\frac{\partial F}{\partial p_j}\big(p(t),\, z(t),\, \mathrm{x}(t)\big),$$

where to obtain the second equality we use equations $p^i(t) = u_{xi}\big(\mathrm{x}(t)\big)$ and

$$\dot{x}^j(t) = \frac{\partial F}{\partial p_j}\big(p(t),\, z(t),\, \mathrm{x}(t)\big).$$

Writing equations $\dot{x}^j(t) = \dfrac{\partial F}{\partial p_j}\big(p(t),\, z(t),\, \mathrm{x}(t)\big),$

$$\dot{p}^i(t) = -\frac{\partial F}{\partial x_i}\big(p(t),\, z(t),\, \mathrm{x}(t)\big) - \frac{\partial F}{\partial z}\big(p(t),\, z(t),\, \mathrm{x}(t)\big)\, p^i(t),$$

and

$$\dot{z}(t) = \sum_{j=1}^{n} \frac{\partial u}{\partial x_j}\big(\mathrm{x}(t)\big)\, \dot{x}^j(t) = \sum_{j=1}^{n} p^j(t)\frac{\partial F}{\partial p_j}\big(p(t),\, z(t),\, \mathrm{x}(t)\big),$$

in vector notation we have a system of $2n+1$ first-order ODEs called the characteristic equations of the PDE:

$$\dot{p}(t) = -D_x F\,(p(t),\, z(t),\, x(t)) - D_z F\,(p(t),\, z(t),\, x(t))\, p(t)$$
$$\dot{z}(t) = D_p F\,(p(t),\, z(t),\, x(t)) \cdot p(t)$$
$$\dot{x}(t) = D_p F\,(p(t),\, z(t),\, x(t)).$$

We assumed that equation $\dot{x}(t) = D_p F\,(p(t),\, z(t),\, \mathrm{x}(t))$ holds. Notice that in the linear case we got only equations and it turns out that in linear (and quasi-linear) cases, the n-equations given in are redundant.

Variational Principles

In the previous topic, we discovered the characteristic equations which enable us to evolve initial data given to us along some n−1-dimensional surface into an n-dimensional solution surface. This procedure is local in the sense that we need to make sure that the ODEs can be solved with no conflicts or blow ups.

An alternative approach when considering physical systems, is to look at the entire (infinite dimensional) space of possible "paths" a system might take. The "true" path is usually a path that minimizes a certain quantity (by the principle of least action) and we will see that it satisfies certain equations.

A typical example is finding a geodesic between two points A and B on a manifold. In such a case, the quantity for which we seek a minimum is the distance functional, and the set on which we seek the minimum is the set of all continuous (smooth) paths on the manifold, between A and B.

Calculus of Variations

We start with a function:

$$L : \mathbb{R}^n \times \mathbb{R}^n \to \mathbb{R},$$

where $L = L(q, x)$ is called the Lagrangian. We now define the action functional:

$$J[w(\cdot)] := \int_0^t L(\dot{w}(s), w(s))\, ds,$$

defined for functions $w(\cdot) = (w^1(\cdot), \ldots, w^n(\cdot))$ belonging to the admissible class:

$$A := \{ \text{smooth curves w with } w(0) = y,\ w(t) = x \}.$$

We seek a curve $x \in A$ that minimizes J over A. The following theorem can be shown:

Theorem: (Euler-Lagrange equations). The minimizer $x(\cdot)$ solves the system of Euler-Lagrange equations

$$\frac{d}{ds}\left(\frac{\partial L}{\partial q}(\dot{x}(s), x(s)) \right) - \frac{\partial L}{\partial x}(\dot{x}(s), x(s)) = 0$$

for each $o \le s \le t$. These are n coupled second order PDEs.

Hamilton-Jacobi Equations

Characteristics for the General Hamilton-Jacobi Equation

The general Hamilton-Jacobi equation is

$$G(Du, ut, u, x, t) = u_t + H(Du, x) = 0.$$

The function H is called the Hamiltonian.

Denoting $q = (p, p_{n+1})$ and $y = (x, t)$ we have

$$G(q, z, y) = p_{n+1} + H(p, x).$$

Thus, we have the following partial derivatives:

$$D_q G = (D_p H(p, x), 1)$$
$$D_y G = (D_x H(p, x), 0)$$
$$D_z G = 0.$$

Plugging into equation $\dot{x}(t) = D_p F(p(t), z(t), x(t))$, we get

$$\begin{cases} \dot{x}^i(s) = \dfrac{\partial H}{\partial p_i}(p(s),\ x(s)),\ (i = 1, \ldots, n) \\[2mm] \dot{x}^{n+1}(s) = 1 \end{cases}$$

so that the parameter s actually identifies with the time t. Equation
$\dot{p}(t) = -D_x F\left(p(t), z(t), x(t)\right) - D_z F\left(p(t), z(t), x(t)\right) p(t)$ becomes

$$\begin{cases} \dot{p}^i(s)\dfrac{\partial H}{\partial x_i} = (p(s),\ \mathrm{x}(s)),\ (i = 1, \ldots, n) \\[2mm] \dot{p}^{n+1}(s) = 0 \end{cases}$$

and equation $\dot{z}(t) = D_p F\left(p(t), z(t), x(t)\right) \cdot p(t)$ becomes

$$\begin{aligned} \dot{z}(s) &= D_p H\left(p(s),\ \mathrm{x}(s)\right) \cdot p(s) + p^{n+1}(s) \\ &= D_p H\left(p(s),\ \mathrm{x}(s)\right) \cdot p(s) - H\left(p(s),\ x(s)\right) \end{aligned}$$

The first and third of the characteristic equations may be written in vector form

$$\dot{x} = D_p H(p, \mathrm{x})$$
$$\dot{p} = D_x H(p, \mathrm{x})$$

and are called Hamilton's equations. *(Here $x(\cdot) = (x^1(\cdot), \ldots, x^n(\cdot))$, $p(\cdot) = (p^1(\cdot), \ldots, p^n(\cdot))$.)*
The equation for z becomes trivial once we have Hamilton's equations.

Relating Hamilton's Equations with the Euler-Lagrange Equations

We begin by defining the generalized momentum corresponding to the position $\mathrm{x}(\cdot)$ and the velocity $\dot{\mathrm{x}}(\cdot)$ by

$$p(s) := D_q L\left(\dot{\mathrm{x}}(s), x(s)\right)$$

where $0 \le s \le t$. We make the following assumption:

"Convexity assumption": For all $x, p \in \mathbb{R}^n$ the equation $p = D_q L(q, x)$ can be uniquely solved for q as a smooth function of p and $x, q = q(p, x)$.

Definition (Hamiltonian): The Hamiltonian H associated with the Lagrangian L is

$$H(p, x) := p \cdot q(p, x) - L(q(p, x), x)$$

where $p, x \in \mathbb{R}^n$ and the function q(p, x) is defined implicitly by the assumption above.

The following theorem, whose proof relies on the Euler-Lagrange equations,

Theorem: The functions $\mathrm{x}(\cdot)$ and $\mathrm{p}(\cdot)$ satisfy Hamilton's equations

$$\dot{x} = D_p H(\mathrm{p}, \mathrm{x})$$
$$\dot{p} = D_x H(\mathrm{p}, \mathrm{x})$$

For $0 \le s \le t$. Furthermore, the mapping:

$$s \mapsto H(p(s), x(s))$$

is constant.

We consider the Lagrangian

$$L(q, x) = \frac{1}{2}m|q|^2 - V(x)$$

where m > 0.

Then we have:

$$\frac{\partial L}{\partial q} = mq$$

$$\frac{\partial L}{\partial x} = -DV(x).$$

Thus, the Euler-Lagrange equation $\dfrac{d}{ds}\left(\dfrac{\partial L}{\partial q}(\dot{x}(s), x(s))\right) - \dfrac{\partial L}{\partial x}(\dot{x}(s), x(s)) = 0$

become:

$$\frac{d}{ds}\left(\frac{\partial L}{\partial q} = (\dot{x}(s), x(s))\right) - \frac{\partial L}{\partial x}(\dot{x}(s), x(s)) = \frac{d}{ds}\left(m\dot{x}(s)\right) + DV(x(s))$$

$$= m\ddot{x}(s) - F(x(s)) = 0$$

where $F := -DV$. This is precisely Newton's second law, describing the motion of a particle with mass m in a force field F generated by a potential V.

We check if the assumption above holds, namely, if the equation $p = D_q L(q, x)$ can be uniquely solved for q as a smooth function of p and x. We have:

$$p = D_q L(q, x) = mq$$

So that:

$$q(p, x) = q = \frac{p}{m}.$$

Now, recalling Definition (of the Hamiltonian H associated to L), we have:

$$H(p, x) = p \cdot q(p, x) - L(q(p, x), x)$$

$$= p \cdot \frac{p}{m} - \frac{1}{2}m\left|\frac{p}{m}\right|^2 + V(x)$$

$$= \frac{1}{2m}|p|^2 + V(x),$$

which is the total energy (kinetic+potantial). By Theorem the energy is constant for a solution.

The Initial Value Problem

The initial value problem for the Hamilton-Jacobi equation is the system:

$$\begin{cases} u_t + H(Du) = 0 & \text{in } \mathbb{R}^n \times (0,\infty) \\ u = g & \text{on } \mathbb{R}^n \times \{t = 0\}, \end{cases}$$

where $u = u(x,t) : \mathbb{R}^n \times [0,\infty) \to \mathbb{R}$, $Du = D_x u = (u_{x_1}, \ldots, u_{x_n})$, and the Hamiltonian H and the initial function $g : \mathbb{R}^n \to \mathbb{R}$ are given.

Non-homogeneous Linear Partial Differential Equations

A linear partial differential equation is non-homogeneous if it contains a term that does not depend on the dependent variable. For example, consider the wave equation with a source:

$$u_{tt} = c^2 u_{xx} + s(x,t)$$

$$\text{boundary conditions } u(0,t) = u(L,t) = 0$$

$$\text{initial conditions } u(x,0) = f(x), \, ut(x,0) = g(x)$$

To solve this, we first look for a particular solution v(x, t) of the PDE and boundary conditions. Then the general solution will be $u(x,t) = v(x,t) + w(x,t)$, where $w(x,t)$ is the general solution of the homogeneous PDE $u_{tt} = c^2 u_{xx}$ and boundary conditions. To satisfy our initial conditions, we must take the initial conditions for w as $w(x,0) = f(x) - v(x,0)$, $w_t(x,0) = g(x) - v_t(x,0)$.

Case: Steady State.

If the source term s(x, t) does not depend on the time t (so we can write s(x, t) = s(x)), then we can look for v(x, t) = v(x) not depending on the time t. The PDE becomes $0 = c^2 v'' + s(x)$, and we must solve this subject to the boundary conditions $v(0) = v(L) = 0$. In this case it can be solved by integrating twice. For example, consider the problem:

$$u_{tt} = u_{xx} + x$$

$$\text{boundary conditions } u(0,t) = u(1,t) = 0$$

$$\text{initial conditions } u(x,\ 0) = 0, \quad u_t(x,0) = 1$$

The differential equation says $v'' = -x$. One integration gives $v' = -x^2 / 2 + A$ where A is a constant, another gives $v = -x^3 / 6 + Ax + B$. For $v(0) = 0$ we need $B = 0$, and then for $v(1) = 0$ we need $-1/6 + A = 0$ or $A = 1/6$. So $v(x) = (x - x^3)/6$ is our particular solution.

The other part of the solution, $w(x,t)$, satisfies:

$$w_{tt} = w_{xx}$$

boundary conditions $w(0,t) = w(1,t) = 0$

initial conditions $w(x,\ 0) = -(x - x^3)/6,\ w_t(x,0) = 1$

We could use a Fourier series for this : $w(x,t) = \sum_{n=1}^{\infty} \sin(n\pi x)\left(b_n \cos(n\pi t) + b_n^* \sin((n\pi t))\right)$

where

$$b_n = -\frac{1}{3}\int_0^1 (x - x^3)\sin(n\pi x)\,dx = \frac{2(-1)^n}{n^3\pi^3}$$

$$b_n^* = \frac{2}{n\pi}\int_0^1 \sin(n\pi x)\,dx = \frac{2(1 - (-1)^n)}{n^2\pi^2}$$

And thus the complete solution is:

$$u(x,t) = \frac{x - x^3}{6} + \sum_{n=1}^{\infty}\left(\frac{2(-1)^n}{n^3\pi^3}\cos(n\pi t) + \frac{2(1 - (-1)^n)}{n^2\pi^2}\sin(n\pi t)\right)\sin(n\pi x)$$

Case: Exponential in t.

If the source term is a function of x times an exponential in t, we may look for a particular solution v(x,t) that is also of this form. For example, consider:

$$u_{tt} = u_{xx} + xe^{-t}$$

boundary conditions $u(0,t) = u(1,t) = 0$

initial conditions $u(x,0) = 0,\quad u_t(x,0) = 1$

We look for a solution of the form $v(x,t) = V(x)e^{-t}$. Then the differential equation says $V(x)e^{-t} = V''(x)e^{-t} + xe^{-t}$ or $V = V'' + x$. The general solution of this differential equation is $V(x) = x + c_1 e^x + c_2 e^{-x}$. The boundary conditions say $V(0) = 0 = c_1 + c_2$ and $V(1) = 0 = 1 + c_1 e + c_2 e^{-1}$. Solving for c_1 and c_2 we get $c_1 = -e/(e^2 - 1)$, $c_2 = e/(e^2 - 1)$, i.e.

$$v(x,t) = \left(x - \frac{e^{1+x}}{e^2 - 1} + \frac{e^{1-x}}{e^2 - 1}\right)e^{-t}$$

The initial conditions for w(x, t) are:

$$w(x,\ 0) = -v(x,\ 0) = -V(x) = -x + \frac{e^{1+x} - e^{1-x}}{e^2 - 1}$$

$$wt(x,\ 0) = 1 - v_t(x,0) = 1 + V(x) = 1 + x - \frac{e^{1+x} - e^{1-x}}{e^2 - 1}$$

Case: Arbitrary function of x and t.

Suppose the solutions of the homogeneous equation involve series (such as Fourier sine or cosine series) in functions $\varphi_n(x)$ (what we'll call an eigenfunction expansion): a more-or-less arbitrary function of x can be expanded in such a series. We can write u(x, t) and s(x, t) for any t as such a series, obtaining series expansions where the coefficients are functions of t:

$$u(x,t) = \sum_{n=1}^{\infty} b_n(t)\phi_n(x)$$

$$s(x,t) = \sum_{n=1}^{\infty} c_n(t)\phi_n(x)$$

Our PDE will give us relations between these, which will be ordinary differential equations in $b_n(t)$ for each n. For example, consider the problem:

$$u_{tt} = u_{xx} + xt$$

$$\text{boundary conditions } u(0,t) = u(1,t) = 0$$

$$\text{initial conditions } u(x,0) = 0, \ u_t(x,0) = 1$$

The appropriate eigenfunctions for the homogeneous problem are $\varphi_n(x) = \sin(n\pi x)$, the expansion being the Fourier sine series on the interval [0, 1]. In particular:

$$xt = \sum_{n=1}^{\infty} \frac{2(-1)^{n+1} t}{n\pi} \sin(n\pi t)$$

So $c_n(t) = 2(-1)^{n+1} t / (n\pi)$. Putting these series into the differential equation, we get:

$$\sum_{n=1}^{\infty} b_n''(t)\sin(n\pi x) = -\sum_{n=1}^{\infty} b_n(t)(n\pi)^2 \sin(n\pi x) + \sum_{n=1}^{\infty} c_n(t)\sin(n\pi x).$$

By the uniqueness of Fourier series, the coefficients for each n must match, i.e.

$$b_n'' = (n\pi)^2 b_n + c_n(t) = -(n\pi)^2 b_n - \frac{2(-1)^n t}{n\pi}$$

The initial conditions for u and u_t give us initial conditions for b_n and

$$b_n' : u(x,0) = 0 \text{ so } b_n(0) = 0, \text{ and } u_t(x,0) = 1 = \sum_{n=1}^{\infty} \frac{2(1-(-1)^n}{n\pi} \sin(n\pi x) \text{ so } b_n'(0) = \frac{2(1-(-1)^n}{n\pi}.$$

The general solution of the differential equation for b_n is.

$$b_n(t) = A_n \cos(n\pi t) + B_n \sin(n\pi t) - \frac{2(-1)^n t}{(n\pi)^3}$$

From the initial conditions we get,

$$b_n(0) = 0 = A_n \text{ and } b_n'(0) = \frac{2(1-(-1)^n)}{n\pi} = n\pi B_n - \frac{2(-1)^n}{(n\pi)^3}, \text{ so } B_n = \frac{2(-1)^n}{(n\pi)^4} + \frac{2(1-(-1)^n)}{(n\pi)^2}.$$

The complete solution is:

$$u(x,t) = \sum_{n=1}^{\infty} \left[\frac{2(-1)^n}{(n\pi)^4} + \frac{2(1-(-1)^n)}{(n\pi)^2} \sin(n\pi t) - \frac{2(-1)^n t}{(n\pi)^3} \right] \sin(n\pi x).$$

Second Order Partial Differential Equations

The general second order partial differential equations in two variables is of the form:

$$F\left(x, y, u, \frac{\partial u}{\partial x}, \frac{\partial u}{\partial y}, \frac{\partial^2 u}{\partial x^2}, \frac{\partial^2 u}{\partial x \partial y}, \frac{\partial^2 u}{\partial y^2}\right) = 0.$$

The equation is quasi-linear if it is linear in the highest order derivatives (second order), that is if it is of the form:

$$a(x, y, u, u_x, u_y)u_{xx} + 2b(x, y, u, u_x, u_y)u_{xy} + c(x, y, u, u_x, u_y)u_{yy} = d(x, y, u, u_x, u_y)$$

We say that the equation is semi-linear if the coefficients a, b, c are independent of u. That is if it takes the form:

$$a(x, y))u_{xx} + 2b(x, y)u_{xy} + c(x, y)u_{yy} = d(x, y, u, u_x, u_y)$$

Finally, if the equation is semi-linear and d is a linear function of u, u_x and u_y we say that the equation is linear. That is, when F is linear in u and all its derivatives.

We will consider the semi-linear equation above and attempt a change of variable to obtain a more convenient form for the equation.

Let $\xi = \phi(x, y), \eta = \psi(x, y)$ be an invertible transformation of coordinates. That is,

$$\frac{\partial(\xi, \eta)}{\partial(x, y)} \begin{vmatrix} \dfrac{\partial \phi}{\partial x} & \dfrac{\partial \phi}{\partial y} \\ \dfrac{\partial \psi}{\partial x} & \dfrac{\partial \psi}{\partial y} \end{vmatrix} \neq 0.$$

By the chain rule:

$$u_x = u_\xi \phi_x + u_\eta \psi_x, \ u_y = u_\xi \phi_y + u_\eta \psi_y$$

$$u_{xx} = u_\xi \phi_{xx} + \phi_x \left(u_{\xi\xi}\phi_x + u_{\xi\eta}\psi_x \right) + u_\eta \psi_{xx} + \psi_x \left(u_{\eta\xi}\phi_x + u_{\eta\eta}\psi_x \right)$$
$$= u_{\xi\xi}\phi_x^2 + 2u_{\xi\eta}\phi_x\psi_x + u_{\eta\eta}\psi_x^2 + \text{first order derivatives of } u$$

Similarly,

$$u_{yy} = u_{\xi\xi}\phi_y^2 + 2u_{\xi\eta}\phi_y\psi_y + u_{\eta\eta}\psi_y^2 + \text{first order derivatives of } u$$
$$u_{xy} = u_{\xi\xi}\phi_x\psi_y + u_{\xi\eta}\left(\phi_x\psi_y + \phi_y\psi_x \right) + u_{\eta\eta}\psi_x\psi_y + \text{first order derivatives of } u$$

Substituting into the partial differential equation we obtain,

$$A(\xi,n)u_{\xi\xi} + 2B(\xi,\eta)u_{\xi\eta} + C(\xi,\eta)u_{\eta\eta} = D\left(\xi, \eta, u, u_\xi, u_\eta \right)$$

where,

$$A(\xi,n) = a\phi_x^2 + 2b\phi_x\phi_y + c\phi_y^2$$
$$B(\xi,n) = a\phi_x\psi_x + b\left(\phi_x\psi_y + \psi_x\phi_y \right) + c\phi_y\psi_y$$
$$C(\xi,n) = a\psi_x^2 + 2\psi_x\psi_y + c\psi_y^2.$$

It easily follows that:

$$B^2 - AC = \left(b^2 - ac\right)\left(\frac{\partial(\xi,\eta)}{(x,y)} \right)^2.$$

Therefore $B^2 - AC$ has the same sign as $b^2 - ac$. We will now choose the new coordinates $\xi = \phi(x,y)$, $\eta = \psi(x, y)$ to simplify the partial differential equation.

$\phi(x,y) = \text{constant}$, $\psi(x,y) = \text{constant}$ defines two families of curves in R². On a member of the family $\phi(x,y) = \text{constant}$, we have that:

$$\frac{d\phi}{dx} = \phi_x + \phi_y y' = 0.$$

Therefore substituting in the expression for $A(\xi,\eta)$ we obtain:

$$A(\xi,\eta) = a\phi_y^2 y'^2 - 2b\phi_y^2 y' + c\phi_y^2$$
$$= \phi_y^2 \left[ay'^2 - 2by' + c \right]$$

We choose the two families of curves given by the two families of solutions of the ordinary differential equation:

$$ay'^2 - 2by' + c = 0.$$

This nonlinear ordinary differential equation is called the characteristic equation of the partial differential equation and provided that $a \neq 0$, $b^2 - ac > 0$ it can be written as:

$$y' = \frac{b \pm \sqrt{b^2 - ac}}{a}$$

For this choice of coordinates $A(\xi, \eta) = 0$ and similarly it can be shown that $C(\xi, \eta) = 0$ also. The partial differential equation becomes:

$$2B(\xi, \eta)u_{\xi\eta} = D(\xi, \eta, u, u_\xi, u_\eta)$$

where it is easy to show that $B(\xi, \eta) \neq 0$. Finally, we can write the partial differential equation in the normal form:

$$u_{\xi\eta} = D(\xi, \eta, u, u_\xi, u_\eta).$$

The two families of curves $\phi(x, y) = $ constant, $\psi(x, y) = $ constant obtained as solutions of the characteristic equation are called characteristics and the semi-linear partial differential equation is called hyperbolic if b² − ac > 0 whence it has two families of characteristics and a normal form as given above.

If $b^2 = ac < 0$, then the characteristic equation has complex solutions and there are no real characteristics. The functions $\phi(x, y)$, $\psi(x, y)$ are now complex conjugates. A change of variable to the real coordinates:

$$\xi = \phi(x, y) + \psi(x, y), \; \eta = -i(\phi(x, y) - \psi(x, y))$$

results in the partial differential equation where the mixed derivative term vanishes,

$$u_{\xi\xi} + u_{\eta\eta} = D(\xi, \eta, u, u_\xi, u_\eta).$$

In this case the semi-linear partial differential equation is called elliptic If $b^2 = ac < 0$, Notice that the left hand side of the normal form is the Laplacian. Thus Laplaces equation is a special case of an elliptic equation (with D = 0).

If $b^2 - ac = 0$, the characteristic equation $y' = \dfrac{b}{a}$ has only one family of solutions $\psi(x, y) = $ constant. We make the change of variable:

$$\xi = x, \; \eta = \psi(x, y).$$

Then,

$$A(\xi, n) = a$$
$$B(\xi, n) = a\psi_x + b\psi_y$$
$$C(\xi, \eta) = a\psi_x^2 + 2b\psi_x\psi_y + c\psi_y^2 = \frac{(a\psi_x + b\psi_y)^2 - (b^2 - ac)\psi_y^2}{a} = \frac{B(\xi, \eta)^2}{a}$$

Also since $\psi(x,y) = \text{constant}$,

$$0 = \psi_x + \psi_y y' = \psi_x = \psi_y \frac{b}{a} = \frac{a\psi_x + b\psi_y}{a} = \frac{B(\xi,\eta)^2}{a}$$

Therefore $B(\xi,n) = 0, C(\xi,\eta) = 0, A(\xi,n) \neq 0$ and the normal form in the case $b^2 - ac = 0$ is :

$$A(\xi,\eta)u_{\xi\xi} = D(\xi,\eta,u,u_\xi,u_\eta)$$

or finally,

$$u_{\xi\xi} = D(\xi,\eta,u,u_\xi,u_\eta)$$

The partial differential equation is called parabolic in the case $b^2 - a = 0$. An example of a parabolic partial differential equation is the equation of heat conduction:

$$\frac{\partial u}{\partial t} - k\frac{\partial^2 u}{\partial x^2} = 0 \text{ where } u = u(x,t).$$

Example: Classify the following linear second order partial differential equation and find its general solution.

$$xyu_{xx} + x^2 u_{xy} - yu_x = 0.$$

In this example $b^2 - ac = \left(\frac{x^2}{2}\right)^2 \geq 0 \therefore$ the partial differential equation is hyperbolic provided $x \neq 0$, and parabolic for x = 0.

For x ≠ 0 the characteristic equations are:

$$y' = \frac{b \pm \sqrt{b^2 - ac}}{a} = \frac{\frac{x^2}{2} \pm \frac{x^2}{2}}{xy} = 0 \text{ or } \frac{x}{y}$$

If $y' = 0$, $y = \text{constant}$.

If $y' = \frac{x}{y}$, $x^2 - y^2 = \text{constant}$. Therefore two families of characteristics are:

$$\xi = x^2 - y^2, \eta = y.$$

Using the chain rule a number of times we calculate the partial derivatives:

$$u_x = u_\xi 2x + u_\eta 0 = 2xu_\xi$$

$$u_{xx} = 2u_\xi + 2x(u_{\xi\xi}2x + u_{\xi\eta}0) = 2u_\xi + 4x^2 u_{\xi\xi}$$

$$u_{xy} = 2x(u_{\xi\xi}(-2y) + u_{\xi\eta}1) = -4xyu_{\xi\xi} + 2xu_{\xi\eta}$$

Substituting into the partial differential equation we obtain the normal form:

$$u_{\xi\eta} = 0 \left(\text{provided } x \neq 0\right).$$

Integrating this equation with respect to η:

$$u_{\xi} = f(\xi),$$

where f is an arbitrary function of one real variable. Integrating again with respect to ξ:

$$u(\xi,n) = \int f(\xi)d\xi + G(\eta) = F(\xi) + G(\eta)$$

where F, G are arbitrary functions of one real variable. Reverting to the original coordinates we find the general solution:

$$u(x,y) = F\left(x^2 - y^2\right) + G(y).$$

Example: Classify, reduce to normal form and obtain the general solution of the partial differential equation,

$$x^2 u_{xx} + 2xy u_{xy} + y^2 u_{yy} = 4x^2$$

For this equation $b^2 - ac = (xy)^2 - x^2 y^2 = 0$ $\therefore$ the equation is parabolic everywhere in the plane (x, y). The characteristic equation is:

$$y' = \frac{b}{a} = \frac{xy}{x^2} = \frac{y}{x}.$$

Therefore there is one family of characteristics $\dfrac{y}{x} = \text{constant}$.

Let $\xi = x$ and $\eta = \dfrac{y}{x}$. Then using the chain rule,

$$u_x = u_{\xi}1 + u_{\eta}\left(\frac{-y}{x^2}\right) - u_{\xi} - \frac{y}{x^2}u\eta$$

$$u_y = u_{\xi}0 + u_{\eta}\left(\frac{1}{x}\right) = \frac{1}{x}u_{\eta}$$

$$u_{xx} = u_{\xi\xi}1 + u_{\xi\eta}\left(\frac{-y}{x^2}\right) + \frac{2y}{x^3}u_{\eta} - \frac{y}{x^2}\left(u_{\eta\xi}1 + u_{\eta\eta}\left(\frac{-y}{x^2}\right)\right)$$

$$= u_{\xi\xi} - \frac{2y}{x^2}u_{\xi\eta} + \frac{y^2}{x^4}u_{\eta\eta} + \frac{2y}{x^3}u_{\eta}$$

$$u_{yy} = \frac{1}{x}\left(u_{\eta\xi}0 + u_{\eta\eta}\left(\frac{1}{x}\right)\right) = \frac{1}{x^2}u_{\eta\eta}$$

$$u_{yx} = \frac{1}{x^2}u_{\eta} + \frac{1}{x}\left(u_{\eta\xi}1 + u_{\eta\eta}\left(-\frac{y}{x^2}\right)\right)$$

$$= \frac{1}{x}u_{\xi\eta} - \frac{y}{x^3}u_{\eta\eta} - \frac{1}{x^2}u_{\eta}.$$

Substituting into the partial differential equation we obtain the normal form:

$$u_{\xi\xi} = 4.$$

Integrating with respect to ξ:

$$u_\xi = 4\xi + f(\eta)$$

where f is an arbitrary function of a real variable. Integrating again with respect to ξ:

$$u(\xi,n) = 2\xi^2 + \xi f(\eta) + g(\eta),$$

Therefore the general solution is given by:

$$u(x,y) = 2x^2 + xf\left(\frac{y}{x}\right) + g\left(\frac{y}{x}\right)$$

where f, g are arbitrary functions of a real variable.

The Wave Equation

The wave equation is a linear second-order partial differential equation which describes the propagation of oscillations at a fixed speed in some quantity y:

$$\frac{1}{v^2}\frac{\partial^2 y}{\partial t^2} = \frac{\partial^2 y}{\partial x^2},$$

where v is the velocity of the wave.

The equation is a good description for a wide range of phenomena because it is typically used to model small oscillations about an equilibrium, for which systems can often be well approximated by Hooke's law. Solutions to the wave equation are of course important in fluid dynamics, but also play an important role in electromagnetism, optics, gravitational physics, and heat transfer. Especially important are the solutions to the Fourier transform of the wave equation, which define Fourier series, spherical harmonics, and their generalizations.

Derivation of the Wave Equation

The derivation of the wave equation varies depending on context. A particularly simple physical setting for the derivation is that of small oscillations on a piece of string obeying Hooke's law. Consider the below diagram showing a piece of string displaced by a small amount from equilibrium:

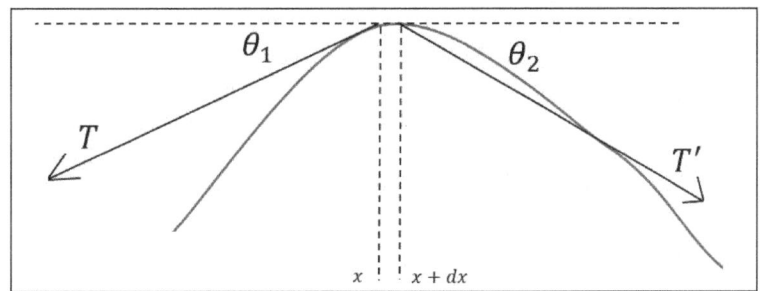

Small oscillations of a string (blue). On a small element of mass contained in a small interval dxdx, tensions T and T′ *pull the element downwards.*

Consider the forces acting on a small element of mass dm contained in a small interval dx. If the displacement is small, the horizontal force is approximately zero. The vertical force is:

$$\sum F_y = -T'\sin\theta_2 - T\sin\theta_1 = (dm)a = \mu dx \frac{\partial^2 y}{\partial t^2},$$

where μ is the mass density $\mu = \frac{\partial m}{\partial x}$ of the string.

On the other hand, since the horizontal force is approximately zero for small displacements, $T\cos\theta_1 \approx T'\cos\theta_2 \approx T$. Therefore,

$$-\frac{\mu dx \frac{\partial^2 y}{\partial t^2}}{T} \approx \frac{T'\sin\theta_2 + T\sin\theta_1}{T} = \frac{T'\sin\theta_2}{T} + \frac{T\sin\theta_1}{T} \approx \frac{T'\sin\theta_2}{T'\cos\theta_2} + \frac{T\sin\theta_1}{T\cos\theta_1} = tan\theta_1 + tan\theta_2.$$

However, $\tan\theta_1 + \tan\theta_2 = -\Delta\frac{\partial y}{\partial x}$, where the difference is between x and $x + dx$. This is because the tangent is equal to the slope geometrically. Dividing over dx, one finds:

$$-\frac{\mu\frac{\partial^2 y}{\partial t^2}}{T} = \frac{\tan\theta_1 + \tan\theta_2}{dx} = -\frac{\Delta\frac{\partial y}{\partial x}}{dx}.$$

The rightmost term above is the definition of the derivative with respect to xx since the difference is over an interval dxdx, and therefore one has:

$$\frac{\mu}{T}\frac{\partial^2 y}{\partial t^2} = \frac{\partial^2 y}{\partial x^2},$$

which is exactly the wave equation in one dimension for velocity $v = \sqrt{\frac{T}{\mu}}$.

Many derivations for physical oscillations are similar. Below, a derivation is given for the wave equation for light which takes an entirely different approach.

Another derivation can be performed providing the assumption that the definition of an entity is the same as the description of an entity.

So, a wave is a squiggly thing, with a speed, and when it moves it does not change shape:

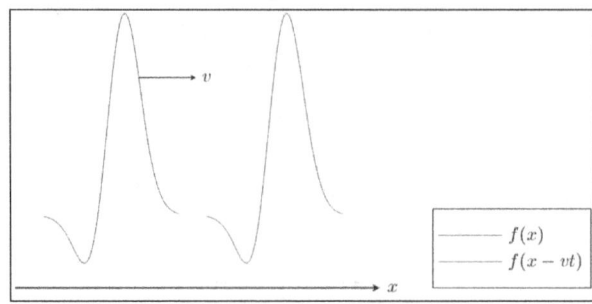

The squiggly thing is $f(x)$, the speed is v, and the red graph is the wave after time t given by a graph transformation of a translation in the x-axis in the positive direction by the distance vt (the distance travelled by the wave travelling at constant speed v over time t): $f(x-vt)$.

Now, since the wave can be translated in either the positive or the negative x direction, let $u=x \pm vt$, so differentiating with respect to x, keeping t constant,

$$\partial u = \partial x,$$

and differentiating with respect to t, keeping x constant,

$$\partial u = \pm v \partial t.$$

So, let us take the second derivative of f with respect to u and substitute the various ∂u:

$$\frac{\partial}{\partial u}\left(\frac{\partial f}{\partial u}\right) = \frac{\partial}{\partial x}\left(\frac{\partial f}{\partial x}\right) = \pm\frac{1}{v}\frac{\partial}{\partial t}\left(\pm\frac{1}{v}\frac{\partial f}{\partial t}\right) \Rightarrow \frac{\partial^2 f}{\partial u^2} = \frac{\partial^2 f}{\partial x^2} = \frac{1}{v^2}\frac{\partial^2 f}{\partial t^2}.$$

So we finally have the wave equation:

$$\frac{\partial^2 f}{\partial x^2} = \frac{1}{v^2}\frac{\partial^2 f}{\partial t^2}.$$

Solution of the Wave Equation

All solutions to the wave equation are superpositions of "left-traveling" and "right-traveling" waves, $f(x + vt)$ and $g(x - vt)$. These are called left-traveling and right-traveling because while the overall shape of the wave remains constant, the wave translates to the left or right in time. Furthermore, any superpositions of solutions to the wave equation are also solutions, because the equation is linear.

Here a brief proof is offered:

Example: Prove that all solutions to the wave equation are superpositions of "left-traveling" and "right-traveling" waves.

Define new coordinates $a = x-vt$ and $b=x + vt$ representing right and left propagation of waves, respectively. Then the partial derivatives can be rewritten as

$$\frac{\partial}{\partial x} = \frac{1}{2}\left(\frac{\partial}{\partial a} + \frac{\partial}{\partial b}\right) \Rightarrow \frac{\partial^2}{\partial x^2} = \frac{1}{4}\left(\frac{\partial^2}{\partial a^2} + 2\frac{\partial^2}{\partial a \partial b} + \frac{\partial^2}{\partial b^2}\right)$$

$$\frac{\partial}{\partial t} = \frac{v}{2}\left(\frac{\partial}{\partial b} - \frac{\partial}{\partial a}\right) \Rightarrow \frac{\partial^2}{\partial t^2} = \frac{v^2}{4}\left(\frac{\partial^2}{\partial a^2} - 2\frac{\partial^2}{\partial a \partial b} + \frac{\partial^2}{\partial b^2}\right).$$

Since the wave equation is:

$$\frac{\partial^2 y}{\partial x^2} - \frac{1}{v^2}\frac{\partial^2 y}{\partial t^2} = 0,$$

substituting in for the partial derivatives yields the equation in the coordinates a and b:

$$\frac{\partial^2 y}{\partial a \partial b} = 0.$$

This is solved in general by $y = f(a) + g(b) = f(x - vt) + g(x + vt)$ as claimed.

The most commonly used examples of solutions are harmonic waves:

$$y(x,t) = A \sin(x - vt) + B \sin(x + vt),$$

where y_0 is the amplitude of the wave and A and B are some constants depending on initial conditions.

If the boundary conditions are such that the solutions take the same value at both endpoints, the solutions can lead to standing waves as seen above. These take the functional form:

$$y(x,t) = y_0 \sin(x - vt) + y_0 \sin(x + vt) = 2y_0 \sin(x)\cos(vt),$$

where y_0 is the amplitude of the wave. The solution has constant amplitude and the spatial part $\sin(x)$ has no time dependence.

Formally, there are two major types of boundary conditions for the wave equation:

- Dirichlet boundary conditions: The amplitude is fixed at the boundary point x^*. This can be written as,

$$y(x^*) = y^*,$$

where y^* is some fixed constant. A more useful way of writing this condition is,

$$\frac{dy}{dt}\bigg|_{x=x^*} = 0,$$

which says that the amplitude at the endpoint does not change over time.

- Neumann boundary conditions: The derivative of the amplitude is specified at the end-points. Often, this derivative is taken to be zero. For an approximately massless (very light) string, this means the endpoints are free rather than fixed; for a pipe with pressure waves inside of it, this means that the endpoints are open to the atmosphere:

$$\frac{dy}{dx}\bigg|_{x=x^*} = 0.$$

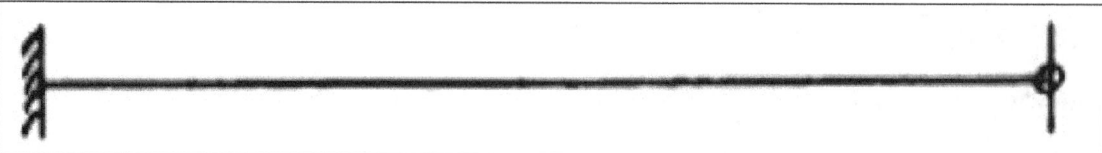

A string with Dirichlet boundary conditions at the left end, where the string is fixed to a wall, and Neumann boundary conditions at the right end, where the string is attached to a freely sliding ring.

Example: A string of tension T and mass density per unit length μ is attached to a small massless ring which slides on a slippery rod. The rod exerts a damping force $F_d = b\dfrac{\partial y}{\partial t}$ on the ring. Find the boundary condition on the oscillations of the string at the end attached to the ring.

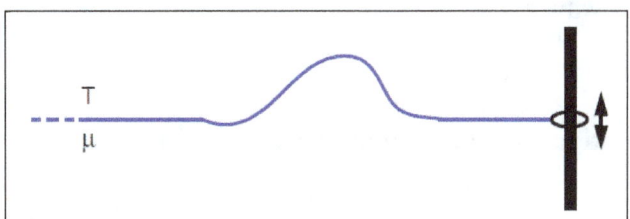

A string attached to a ring sliding on a slippery rod.

The ring is free to slide, so the boundary conditions are Neumann and since the ring is massless the total force on the ring must be zero. Consider the following free body diagram:

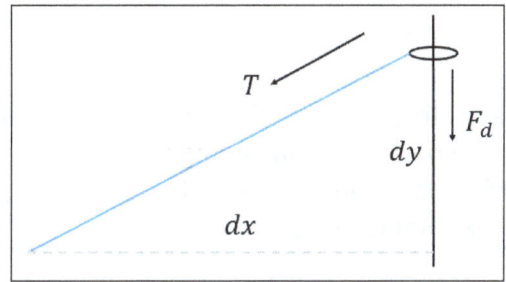

All vertically acting forces on the ring at the end of the oscillating string.

Using the fact that the wave equation holds for small oscillations only, $dx \gg dy$. Balancing the forces in the vertical direction thus yields

$$-T\tfrac{\partial y}{\partial x} - b\tfrac{\partial y}{\partial t} = 0 \;\Rightarrow\; \frac{\partial y}{\partial x} = -\frac{b\partial y}{T\partial x}.$$

This slope condition is the Neumann boundary condition on the oscillations of the string at the end attached to the ring.

One way of writing down solutions to the wave equation generates Fourier series which may be used to represent a function as a sum of sinusoidals. This method uses the fact that the complex exponentials e^{-iwt} are eigenfunctions of the operator $\dfrac{\partial^2}{\partial t^2}$. Using this fact, ansatz a solution for a particular w:

$$y(x,t) = e^{-iwt} f(x),$$

where the exponential has essentially factored out the time dependence. Plugging into the wave equation, one finds:

$$\frac{\partial^2 y}{\partial t^2} = -w^2 y(x,t) = v^2 \frac{\partial^2 y}{\partial x^2} = v^2 e^{-iwt} \frac{\partial^2 f}{\partial x^2}.$$

The function f therefore satisfies the equation:

$$\frac{\partial^2 f}{\partial x^2} = -\frac{w^2}{v^2} f.$$

This is solved by the plane waves:

$$f(x) = f_0 e^{\pm iwx/v}.$$

Therefore, the general solution for a particular ω can be written as:

$$y(x,t) = f_0 e^{i\frac{w}{v}(x \pm vt)}.$$

This is consistent with the assertion above that solutions are written as superpositions of $f(x - vt)$ and $g(x + vt)$ for some functions f and g. By the linearity of the wave equation, an arbitrary solution can be built up in terms of superpositions of the above solutions that have ω fixed. This is exactly the statement of existence of the Fourier series.

Physical Applications

The wave equation governs a wide range of phenomena, including gravitational waves, light waves, sound waves, and even the oscillations of strings in string theory. Depending on the medium and type of wave, the velocity v can mean many different things, e.g. the speed of light, sound speed, or velocity at which string displacements propagate.

Example: Prove that light obeys the wave equation directly from Maxwell's equations.

Begin by taking the curl of Faraday's law and Ampere's law in vacuum:

$$\vec{\nabla} \times (\vec{\nabla} \times \vec{E}) = \frac{\partial}{\partial t} \vec{\nabla} \times \vec{B} = -\mu_0 \in_0 \frac{\partial^2 E}{\partial t^2}$$

$$\vec{\nabla} \times (\vec{\nabla} \times \vec{B}) = \mu_0 \in_0 \frac{\partial}{\partial t} \vec{\nabla} \times \vec{E} = -\mu_0 \in_0 \frac{\partial^2 B}{\partial t^2}.$$

Now using the vector identity:

$$\vec{\nabla} \times (\vec{\nabla} \times A) = \vec{\nabla}(\vec{\nabla} \cdot A) - \vec{\nabla}^2 A,$$

the left-hand sides can also be rewritten. Since $\vec{\nabla} \cdot \vec{E} = \vec{\nabla} \cdot \vec{B} = 0$ according to Gauss' laws for electricity and magnetism in vacuum, this reduces to:

$$\vec{\nabla} \times (\vec{\nabla} \times \vec{E}) = \vec{\nabla}^2 \vec{E}, \quad \vec{\nabla} \times (\vec{\nabla} \times \vec{B}) = -\vec{\nabla}^2 \vec{B}.$$

Equating both sides above gives the two wave equations for $\vec{E}$ and $\vec{B}$:

$$\vec{\nabla}^2 E = \mu_0 \in_0 \frac{\partial^2 E}{\partial t^2}, \quad \vec{\nabla}^2 B = \mu_0 \in_0 \frac{\partial^2 B}{\partial t^2}.$$

Since it can be numerically checked that $c = \dfrac{1}{\sqrt{\mu_0 \, \epsilon_0}}$, this shows that the fields making up light

obeys the wave equation with velocity c as expected.

Example: A plasma is an ionized gas, typically very hot. Ripples in a plasma, in the form of perturbations \rho$\rho$ to the plasma density, satisfy a modified wave equation:

$$v^2 \frac{\partial^2 \rho}{\partial x^2} - w_p^2 \rho = \frac{\partial^2 \rho}{\partial t^2},$$

where v is the speed at which the perturbations propagate and w_p^2 is a constant, the *plasma frequency.*

What is the frequency of traveling wave solutions for small velocities $v \approx 0$?

Ansatz a solution $\rho = \rho_0 e^{i(kx - wt)}$. Plugging in, one finds the equation:

$$-v^2 k^2 \rho - w_p^2 \rho = -w^2 \rho,$$

i.e. the dispersion relation:

$$w^2 = w_p^2 + v^2 k^2 \Rightarrow w = \sqrt{w_p^2 + v^2 k^2}.$$

For small velocities $v \approx 0$, the binomial theorem gives the result:

$$w \approx w_p + \frac{v^2 k^2}{2 w_p}.$$

The size of the plasma frequency w_p thus sets the dynamics of the plasma at low velocities.

Higher-Order Partial Differential Equations

Apart from second-order PDEs, higher-order equations also quite often arise in applications. Below are only a few important examples of such equations with some of their solutions.

Higher-Order Linear Partial Differential Equations

Equation of transverse vibration of elastic rod:

$$\frac{\partial^2 w}{\partial t^2} + a^2 \frac{\partial^4 w}{\partial x^4} = 0.$$

The equation has the following particular solutions:

$$w(x,t) = [A \sin(\lambda x) + B \cos(\lambda x) + C \sin h(\lambda x) + D \cos(\lambda x)] \sin(\lambda^2 a t),$$
$$w(x,t) = [A_1 \sin(\lambda x) + B_1 \cos(\lambda x) + C_1 \sin h(\lambda x) + D_1 \cos(\lambda x)] \cos(\lambda^2 a t),$$

where A, B, C, D , A_1, B_1, C_1, D_1, and λ are arbitrary constants.

Biharmonic equation:

$$\Delta\Delta w = 0,$$

where $\Delta\Delta$ is the biharmonic operator,

$$\Delta\Delta \equiv \Delta^2 = \frac{\partial^4}{\partial x^4} + 2\frac{\partial^4}{\partial x^2 \partial y^2} + \frac{\partial^4}{\partial y^4}.$$

The biharmonic equation is encountered in plane problems of elasticity (w is the Airy stress function). It is also used to describe slow flows of viscous incompressible fluids (w is the stream function).

Various representations of the general solution to equation $\Delta\Delta w = 0$, in terms of harmonic functions include:

$$w(x,y) = xu_1(x,y) + u_2(x,y),$$
$$w(x,y) = yu_1(x,y) + u_2(x,y),$$
$$w(x,y) = (x^2 + y^2)u_1(x,y) + u_2(x,y),$$

Where u_1 and u_2 are arbitrary functions satisfying the Laplace equation $\Delta uk = 0 (k = 1, 2)$.

Complex form of representation of the general solution:

$$w(x,y) = \mathrm{Re}[\bar{z} f(z) + g(z)],$$

where $f(z)$ and $g(z)$ are arbitrary analytic functions of the complex variable $z = x + iy$; $\bar{z} = x - iy, i^2 = -1$. The symbol $\mathrm{Re}[A]$ stands for the real part of a complex quantity A.

Higher-order Nonlinear Partial Differential Equations

Korteweg–de Vries equation:

$$\frac{\partial w}{\partial t} + \frac{\partial^3 w}{\partial x^3} - 6w\frac{\partial w}{\partial x} = 0.$$

It is used in many sections of nonlinear mechanics and theoretical physics for describing one-dimensional nonlinear dispersive non-dissipative waves. In particular, the mathematical modeling of moderate-amplitude shallow-water surface waves is based on this equation.

Equation of a steady laminar boundary layer on a flat plate:

$$\frac{\partial w}{\partial y}\frac{\partial^2 w}{\partial x \partial y} - \frac{\partial w}{\partial x}\frac{\partial^2 w}{\partial y^2} = a\frac{\partial^3 w}{\partial y^3}.$$

where w is the stream function.

Boussinesq equation:

$$\frac{\partial^2 w}{\partial t^2} + \frac{\partial}{\partial x}\left(w\frac{\partial w}{\partial x}\right) + \frac{\partial^4 w}{\partial x^4} = 0.$$

This equation arises in several physical applications: propagation of long waves in shallow water, one-dimensional nonlinear lattice-waves, vibrations in a nonlinear string, and ion sound waves in a plasma.

Equation of motion of a viscous fluid:

$$\frac{\partial w}{\partial y}\frac{\partial}{\partial x}(\Delta w) - \frac{\partial w}{\partial x}\frac{\partial}{\partial y}(\Delta w) = a\Delta\Delta w, \quad \Delta w = \frac{\partial^2 w}{\partial x^2} + \frac{\partial^2 w}{\partial y^2}.$$

This is a two-dimensional stationary equation of motion of a viscous incompressible fluid—it is obtained from the Navier–Stokes equations by the introduction of the stream function w.

The Heat Equation

The heat equation is a partial differential equation which governs the temperature distribution in an object.

In this equation,

$$\frac{\partial T}{\partial t} = \frac{k}{\rho c}\frac{\partial^2 T}{\partial x^2},$$

the temperature T is a function of position x and time t, and k, ρ, and c are respectively, the thermal conductivity, density, and specific heat capacity of the metal, and $k/\rho c$ is called the diffusivity.

Deriving the Heat Equation

The heat equation can be derived from conservation of energy: the time rate of change of the heat stored at a point on the bar is equal to the net flow of heat into that point. This process clearly obeys the continuity equation. If Q is the heat at each point and **V** is the vector field giving the flow of the heat, then:

$$\frac{\partial Q}{\partial t} + \nabla \cdot \mathbf{V} = 0$$

According to the Second Law of Thermodynamics, if two identical bodies are brought into thermal contact and one is hotter than the other, then heat must flow from the hotter body to the colder one at a rate proportional to the temperature difference. Therefore **V** is proportional to the negative gradient of the temperature, so **V**=-k∇T where k is the thermal conductivity of the metal. In one dimension this reduces to **V**=(-k∂T/∂x)**x** where **x** is the unit vector in the

+x-direction. Furthermore, Q=ρcT, so we get the heat equation by plugging in these expressions for **V** and Q:

$$\rho c \frac{\partial \mathbf{V}}{\partial t} + \nabla \cdot (-k \frac{\partial T}{\partial x} \mathbf{x}) = 0$$

$$\therefore \frac{\partial T}{\partial t} - \frac{k}{\rho c} \frac{\partial^2 T}{\partial x^2} = 0$$

Solving the Heat Equation

Before we can go any further, we need to show that there must exist a unique solution to the heat equation for any physically meaningful initial and boundary conditions. The laws of thermodynamics tell us that no matter what the temperature distribution of the bar is initially, the system *must* undergo a process that brings the bar to thermal equilibrium, and this process must obey the heat equation, so solutions to the heat equation exist for physically meaningful initial and boundary conditions. Furthermore, one of the basic assumptions of classical physics is that identical experimental conditions must lead to identical results, so the particular way in which the bar progresses to thermal equilibrium is uniquely specified by the initial and boundary conditions.

This means that if $f(x,t)$ and $g(x,t)$ are two different functions that satisfy the same IBVP for the heat equation, then f and g have the same form. Furthermore the heat equation is linear so if f and g are solutions and α and β are any real numbers, then $\alpha f + \beta g$ is also a solution. So we can conclude that the solution is going to be a linear combination of functions of the same form.

Consider the following function, which we could have guessed by inspection or by trial and error:

$$f(x,t) = \sin\left(\frac{n\pi}{L} x\right) e^{-kn^2\pi^2 / \rho c L^2}$$

Where n is a positive integer greater than zero. This function satisfies the heat equation:

$$\frac{\partial f}{\partial t} = -\frac{k}{\rho c} \frac{n^2 \pi^2}{L^2} \sin\left(\frac{n\pi}{L} x\right) e^{-kn^2\pi^2 t / \rho c L^2}$$

$$\frac{k}{\rho c} \frac{\partial^2 f}{\partial x^2} = -\frac{k}{\rho c} \frac{n^2 \pi^2}{L^2} \sin\left(\frac{n\pi}{L} x\right) e^{-kn^2\pi^2 t / \rho c L^2}$$

$$\therefore \frac{\partial f}{\partial t} = \frac{k}{\rho c} \frac{\partial^2 f}{\partial x^2}$$

This function also satisfies the boundary conditions since sin(0)=sin($n\pi$)=0. Therefore the general solution is:

$$T(x,t) = \sum_{n=1}^{\infty} A_n \sin\left(\frac{n\pi x}{L}\right) e^{-kn^2\pi^2 t / \rho c L^2}$$

The problem will be solved if we can find the coefficients A such that this general solution satisfies the initial condition. That is, we need to find A such that:

$$T(x,0) = \sum_{n=1}^{\infty} A_n \sin\left(\frac{n\pi x}{L}\right)$$

This is called a *Fourier sine series* expansion for the initial conditions. The coefficients A_n called the *Fourier coefficients*.

Computing the Fourier Coefficients

The initial condition $T(x,0)$ is a piecewise continuous function on the interval $[0,L]$ that is zero at the boundaries. It turns out that the set of functions with these properties is a vector space under addition and scalar multiplication. We will call this vector space $\mathbb{S}^L$.

This vector space be equipped with an inner product. For $f,g \in \mathbb{S}^L$, a possible inner product is:

$$\langle f,g \rangle = \frac{2}{L}\int_0^L f(x)g(x)dx$$

An inner product generalizes the idea of the dot product. We can find the components of a geometric vector by projecting it onto the axis using the dot product with the unit vectors, which form a basis for $\mathbb{R}^n$. In the same way, if we can find a basis for $\mathbb{S}^L$ then we can project any $f \in \mathbb{S}^L$ onto the basic functions in order to represent f as a linear combination of basis functions.

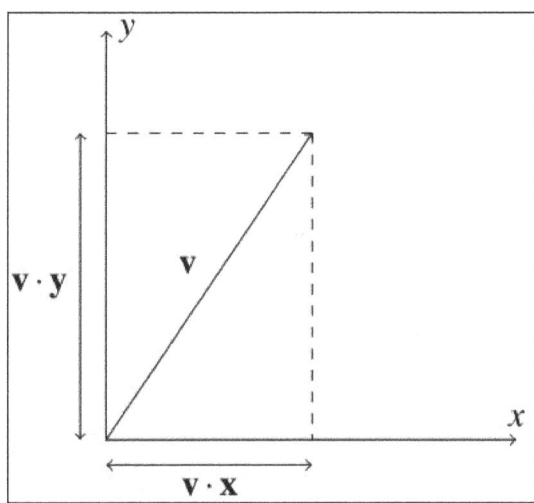

For integers $m, n>0$, functions of the form $\sin(n\pi x/L)$ are orthonormal:

$$\left\langle \sin\left(\frac{m\pi x}{L}\right), \sin\left(\frac{n\pi x}{L}\right) \right\rangle = \begin{cases} 1 \text{ if } m = n \\ 0 \text{ if } m \neq n \end{cases}$$

So we can represent any function $f \in \mathbb{S}^L$ as a linear combination of functions from the basis set:

$$\left\{ \sin\left(\frac{\pi x}{L}\right), \sin\left(\frac{2\pi x}{L}\right), \sin\left(\frac{3\pi x}{L}\right), \sin\left(\frac{4\pi x}{L}\right), ... \right\}$$

And the coefficients of that linear combination will be given by the *Euler integral*:

$$A_n = \frac{2}{L} \int_0^L f(x) \sin\left(\frac{n\pi x}{L}\right) dx$$

To demonstrate, let's find the Fourier coefficients of a unit sawtooth pulse:

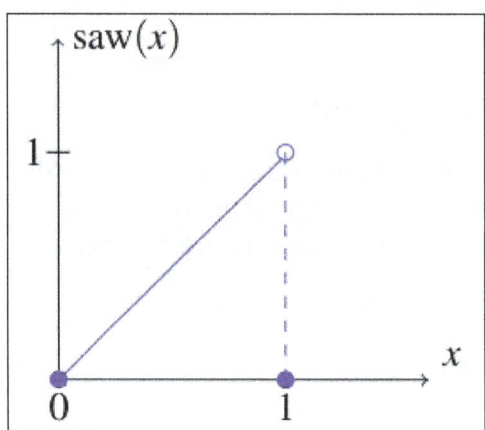

$$\text{saw}(x) = \begin{cases} x & \text{for } 0 \le x < 1 \\ 0 & \text{for } x = 1 \end{cases}$$

Clearly, $\text{saw}(x) \in \mathbb{S}^1$ so:

$$\text{saw}(x) = \sum_{n=1}^{\infty} A_n \sin(n\pi x)$$

and the Fourier coefficients are given by:

$$A_n = \langle \text{saw}(x), \sin(n\pi x) \rangle = 2 \int_0^L x \sin(n\pi x) dx$$

$$= \frac{2}{n\pi}(-1)^{n+1}$$

so the Fourier series expansion for the sawtooth wave is:

$$\text{saw}(x) = \sum_{n=1}^{\infty} \frac{2}{n\pi}(-1)^{n+1} \sin(n\pi x)$$

This animation shows how the Fourier series approaches the sawtooth as the number of sine terms in the sum increases.

The error near x=1 is called the *Gibbs phenomenon*. The Gibbs phenomenon is an unavoidable error that causes the Fourier series of a discontinuous function to overestimate the function's value at a discontinuity by about 9%. The Gibbs phenomenon can never be completely elim-inated, but as the number of terms in the Fourier series approaches infinity, the error con-verges to being localized entirely at the point of discontinuity. So for example if we included an infinite number of terms in the Fourier series expansion for the sawtooth, we would find that the series would be exactly equal to x for $0 \le x < 1$ and at $x=1$ the series would have a value of about 1.09.

What this tells us is that solving the homogeneous IBVP for the heat equation amounts to using the Euler integral to find the Fourier coefficients:

$$T(x,t) = \sum_{n=1}^{\infty} A_n \sin\left(\frac{n\pi x}{L}\right) e^{-kn^2\pi^2 t/\rho c L^2}$$

$$A_n = \frac{2}{L} \int_0^L T(x,0) \sin\left(\frac{n\pi x}{L}\right) dx$$

Example: Bar initially at uniform temperature.

An insulated, meter-long bar with $k/\rho c = 0.1 \text{m}^2/\text{s}$ (unrealistic but easier for graphing purposes) is initially 100 °C throughout its length before cooling elements set to 0 °C are clamped to its ends at $t=0$. The initial and boundary conditions are:

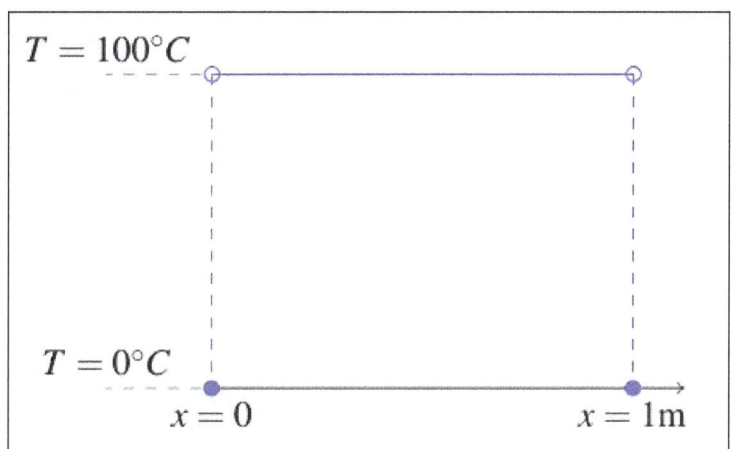

$$\begin{cases} T(x,0) = 100° C & \text{for } 0 < x < 1 \\ T(0,t) = T(1,t) = 0° C & \text{for } 0 \le t \end{cases}$$

The Fourier coefficients are:

$$A_n = \frac{200}{L} \int_0^1 \sin(n\pi x) dx = \frac{200}{\pi n}((-1)^{n+1} + 1)$$

So the solution is:

$$T(x,t) = \sum_{n=1}^{\infty} \frac{200}{\pi n}((-1)^{n+1} + 1) \sin(n\pi x) e^{-kn^2\pi^2 t/\rho c}$$

References

- Partial-Differential-Equation: mathworld.wolfram.com, Retrieved 26 April, 2019

- Wave-equation: brilliant.org, Retrieved 25 February, 2019

- Higher-Order-Partial-Differential-Equations, Partial-differential-equation: scholarpedia.org, Retrieved 03 May, 2019

- The-heat-equation, cantors-paradise: medium.com, Retrieved 02 March, 2019

- Yehuda Pinchover and Ya'akov Rubinstein, Introduction to Partial Differential Equations (Hebrew). Technion, Haifa, Israel, Second Edition, 2003

Numerical Solution of Differential Equations

Numerical solutions of differential equations are used in finding numerical approximations to the solutions of ordinary differential equations. It uses Picard's method, Euler method and Runge-Kutta method to find these approximations. This chapter closely examines these methods of numerical solutions of differential equations to provide an extensive understanding of the subject.

Picard's Method

Picard's Method of solving a differential equation approximately is one of successive approximation; that is, it is an iterative method in which the numerical results become more and more accurate, the more times it is used.

An approximate value of y (taken, at first, to be a constant) is substituted into the right hand side of the differential equation,

$$\frac{dy}{dx} = f(x,y).$$

The equation is then integrated with respect to x giving y in terms of x as a second approximation, into which given numerical values are substituted and the result rounded off to an assigned number of decimal places or significant figures. The iterative process is continued until two consecutive numerical solutions are the same when rounded off to the required number of decimal places.

Notation

Imagine, for example, that you wished to solve the differential equation,

$$\frac{dy}{dx} = 3x^2.$$

Given, $y = y_0 = 7$ when $x = x_0 = 2$.

This of course can be solved exactly to give,

$$y = x^3 + C,$$

which requires that,

$$7 = 2^3 + C.$$

Hence,

$$y - 7 = x^3 - 2^3;$$

or,

$$y - y_0 = x^3 - x_0^3.$$

Thus,

$$\int_{y_0}^{y} dy = \int_{x_0}^{x} 3x^2 dx.$$

In other words,

$$\int_{x_0}^{x} \frac{dy}{dx} dx = \int_{x_0}^{x} 3x^2 dx.$$

The rule, in future, therefore, will be to integrate both sides of the given differential equation with respect to x, from x_0 to x.

Given,

$$\frac{dy}{dx} = x + y^2,$$

and y = 0 when x = 0, determine the value of y when x = 0.3, correct to four places of decimals.

Solution:

To begin the solution, you proceed as follows:

$$\int_{x_0}^{x} \frac{dy}{dx} dx = \int_{x_0}^{x} (x + y^2) dx,$$

where $x_0 = 0$.

Hence,

$$y - y_0 = \int_{x_0}^{x} (x + y^2) dx,$$

where $y_0 = 0$.

That is,

$$y = \int_{0}^{x} (x + y^2) dx.$$

First Iteration

You do not know y in terms of x yet, so we replace y by the constant value y_0 in the function to be integrated.

The result of the first iteration is thus given, at x = 0.3,

$$y_1 = \int_0^x x\,dx = \frac{x^2}{2} \simeq 0.0450.$$

Second Iteration

Now,

$$\frac{dy}{dx} = x + y_1^2 = x + \frac{x^4}{4}.$$

Therefore,

$$\int_0^x \frac{dy}{dx}\,dx = \int_0^x \left(x + \frac{x^4}{4} \right) dx,$$

which gives,

$$y - 0 = \frac{x^2}{2} + \frac{x^5}{20}.$$

The result of the second iteration is thus given by,

$$y_2 = \frac{x^2}{2} + \frac{x^5}{20} \simeq 0.0451$$

at x = 0.3.

Third Iteration

Now,

$$\frac{dy}{dx} = x + y_2^2$$

$$= x + \frac{x^4}{4} + \frac{x^7}{20} + \frac{x^{10}}{400}.$$

Therefore,

$$\int_0^x \frac{dy}{dx}\,dx = \int_0^x \left(x + \frac{x^4}{4} + \frac{x^7}{20} + \frac{x^{10}}{400} \right) dx,$$

which gives,

$$y - 0 = \frac{x^2}{2} + \frac{x^5}{20} + \frac{x^8}{160} + \frac{x^{11}}{4400}.$$

The result of the third iteration is thus given by,

$$y_3 = \frac{x^2}{2} + \frac{x^5}{160} + \frac{x^8}{160} + \frac{x^{11}}{4400} \approx 0.0451 \; at \; x = 0.3.$$

Hence, y = 0.0451, correct to four decimal places, at x = 0.3.

If,

$$\frac{dy}{dx} = 2 - \frac{y}{x}$$

and y = 2 when x = 1, perform three iterations of Picard's method to estimate a value for y when x = 1.2. Work to four places of decimals throughout and state how accurate is the result of the third iteration.

First Iteration

$$\int_{x_0}^{x} \frac{dy}{dx} dx = \int_{x_0}^{x} \left(2 - \frac{y}{x} \right) dx,$$

where $x_0 = 1$.

That is,

$$y = y_0 \int_{x_0}^{x} \left(2 - \frac{y}{x} \right) dx,$$

where $y_0 = 2$.

Hence,

$$y - 2 = \int_{1}^{x} \left(2 - \frac{y}{x} \right) dx.$$

Replacing y by y_0 = 2 in the function being integrated, you have,

$$y - 2 = \int_{1}^{x} \left(2 - \frac{2}{x} \right) dx.$$

Therefore,

$$y = 2 + [2x - 2\ln x]_1^x$$
$$= 2 + 2x - 2\ln x - 2 + 2\ln 1 = 2(x - \ln x).$$

The result of the first iteration is thus given by,

$$y_1 = 2(x - \ln x) \approx 2.0354,$$

when x = 1.2.

Second Iteration

In this case you use

$$\frac{dy}{dx} = 2 - \frac{y_1}{x} = 2 - \frac{2(x - \ln x)}{x} = \frac{2\ln x}{x}.$$

Hence,

$$\int_1^x \frac{dy}{dx} dx = \int_1^x \frac{2\ln x}{x} dx.$$

That is,

$$y - 2 = \left[(\ln x)^2 \right]_1^x = (\ln x)^2.$$

The result of the second iteration is thus given by

$$y_2 = 2 + (\ln x)^2 \approx 2.0332,$$

when x = 1.2.

Third Iteration

Finally, you use

$$\frac{dy}{dx} = 2 - \frac{y_2}{x} = 2 - \frac{2}{x} - \frac{(\ln x)^2}{x}.$$

Hence,

$$\int_1^x \frac{dy}{dx} dx = \int_1^x \left[2 - \frac{2}{x} - \frac{(\ln x)^2}{x} \right] dx.$$

That is,

$$y - 2 = \left[2x - 2\ln x - \frac{(\ln x)^3}{3} \right]_1^x$$

$$= 2x - 2\ln x - \frac{(\ln x)^3}{3} - 2.$$

The result of the third iteration is thus given by

$$y_3 = 2x - 2\ln x - \frac{(\ln x)^3}{3} \approx 2.0293,$$

when x = 1.2.

The results of the last two iterations are identical when rounded off to two places of decimals, namely 2.03. Hence, the accuracy of the third iteration is two decimal place accuracy.

Euler Method

Euler's method is used for approximating solutions to certain differential equations and works by approximating a solution curve with line segments.

In some cases, it's not possible to write down an equation for a curve, but you can still find approximate coordinates for points along the curve by using simple lines. These line segments have the same slope as the curve, so they stay relatively close to it.

Euler's method is useful because differential equations appear frequently in physics, chemistry and economics, but usually cannot be solved explicitly, requiring their solutions to be approximated. For example, Euler's method can be used to approximate the path of an object falling through a viscous fluid, the rate of a reaction over time, the flow of traffic on a busy road, to name a few.

Suppose you want to find approximate values for the solution of the differential equation,

$$y' = f(t,y),$$

with initial condition,

$$y(t_0) = y_0.$$

The differential equation tells us that the instantaneous rate of change of y at time t can be calculated in terms of y and t alone. Now if Δt is any small interval of time you know that as a first order approximation you can write,

$$y(t + \Delta t) = y(t) + \Delta t \cdot y'(t),$$

in the sense that the difference between the right and left sides, for a fixed value of t, is essentially of order $(Dt)^2$. If y is a solution of the differential equation, this tells us that,

$$y(t + \Delta t) = y(t) + \Delta t \cdot f\left(t; y(t)\right).$$

If $t_1 = t_0$ + Dt you know therefore that,

$$y(t_1) = y_1 = y(t_0) + \Delta t \cdot f(t_0 \cdot y(t_0)) = y_0 + \Delta t \cdot f(t_0 y_0).$$

You can apply this reasoning once again. If $t_2 = t_1 + \Delta t = t_0 + 2\Delta t$ you obtain a further estimate

$$y(t_2) = y_2 = y_1 + \Delta t \cdot f\left(t_1, y_1\right).$$

Etc. If $t_{n+1} = t_n + \Delta t = t_0 + (n+1)\Delta t$ you get an estimate y_{n+1} for $y(t_{n+1})$ from the estimate y_n for $y(t_n)$:

$$y_{n+1} = y_n + \Delta t \cdot f(t_n, y_n).$$

This technique for finding approximate values for the solution of a first order differential equation is the simplest of several similar ones. Each of them proceeds in this stepwise fashion, obtaining an estimate for y(t + Dt) from that for y(t). This one is called Euler's method. It has some great virtues.

It is simple to carry out. Doing it by hand is not impossible (although tedious and hence error prone).

It can be easily implemented on a computer, for example with a spread sheet.

- The step from one estimate to the next is intuitive.

The calculations are easy to set up in a spreadsheet, equally easy to program, not even too bad to do a few steps by hand. Here is a sample run for the equation and initial conditions,

$$y' = y - t, \quad y(0) = 0.5.$$

with step size Dt = 0.1, in the range $|0,1|$.

t	y	$f(t,y)$
0.000000	0.500000	0.500000
0.100000	0.550000	0.450000
0.200000	0.595000	0.395000
0.300000	0.634500	0.334500
0.400000	0.667950	0.267949
0.500000	0.694744	0.194745
0.600000	0.714220	0.114220
0.700000	0.725641	0.025641
0.800000	0.728206	-0.071794
0.900000	0.721026	-0.178974
1.000000	0.703129	-0.296871

Euler's Method and Slope Field

Euler's method has a simple geometric interpretation. Solving a differential equation,

$$y' = f(t,y).$$

Geometrically, finding the graph of a function y = y(t) whose slope at any point (t, y) is equal to f (t, y). We can picture this by drawing a number of small 'slope segments' in the (t, y) plane, the segment at (t, y) having slope f (t, y). These segments suggest to the eye what solutions of the differential equation look like. Euler's method has a direct interpretation in terms of such slope fields. In the figure below we are looking at the differential equation & initial condition,

$$y' = y - t, \quad y(0) = 0.5.$$

This is a linear equation and has the explicit solution,

$$y = (t+1) - 0.5e^t.$$

The differential equation tells us how to draw the slope field and the initial condition tells us where the graph star.

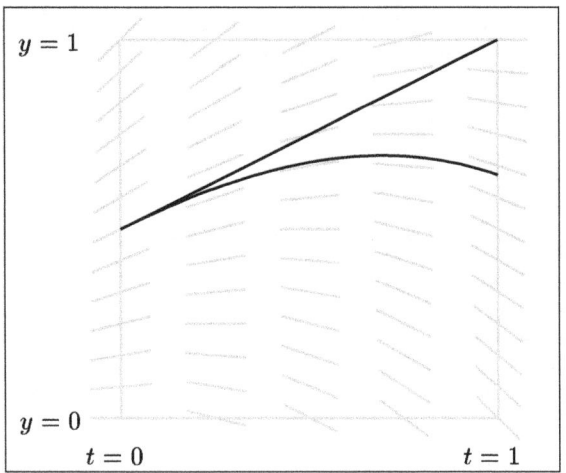 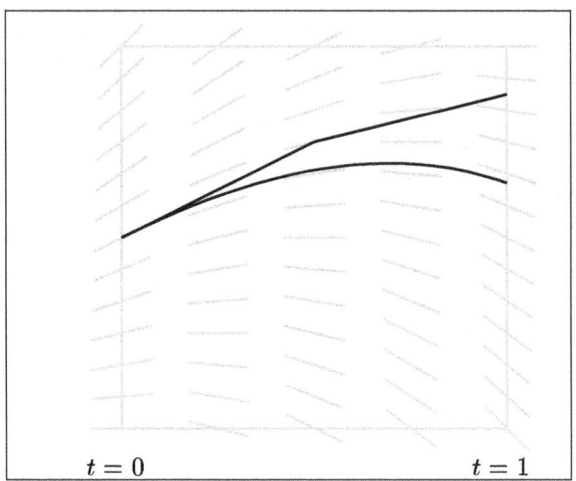

At the starting point you know that the slope of the graph is equal to f (0, 0.5) = 0.5. This means that the tangent line to the graph at the starting point is the line y_* (t)=0.5+0.5 t. This tangent line and the graph of y(t) will lie very close to each other for small values of t and you can then use y_* as an estimate for y(t) for those values of t. After a while, however, it will cease to be a useful approximation. How can you get a better one? You pick some interval Dt and at t = Dt we modify the tangent line. You must modify it by using only information available to us in the calculation so far, and in view of this it seems reasonable to use as the new approximation the straight line through (Dt, y_*(Dt)) whose slope agrees with the slope field at that point.

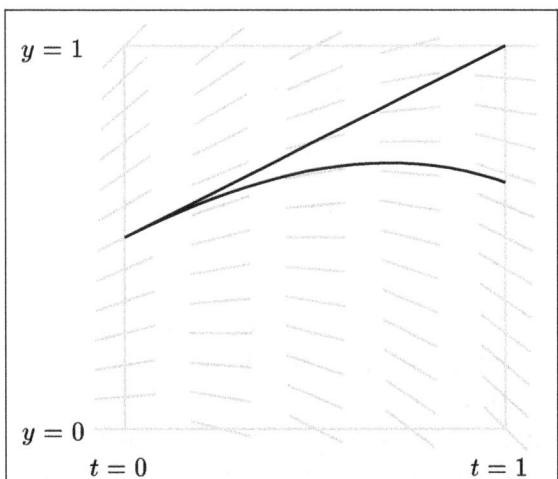 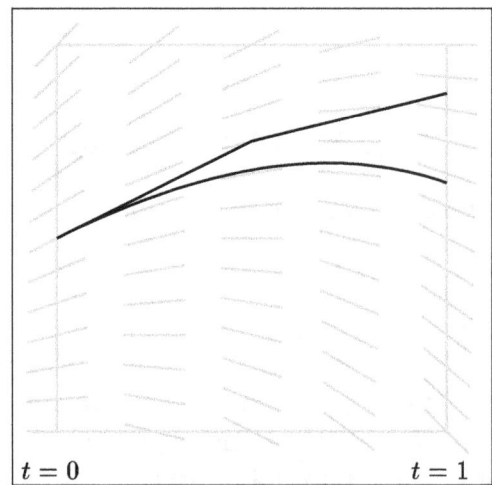

You then make new breaks in the approximation after every interval of size Dt. As you take smaller and smaller values of Dt we get better approximations to the graph of the solution.

The Error in Euler's Method

You expect that the error in the approximation you get from Euler's method decreases as you let the step size Dt get smaller. You can get a good idea of how the error depends on the choice of Dt by plotting in one picture the approximations you get for different values of Dt. Here you let Dt = 1, ½, ¼, etc. again with the differential equation,

$$y' = y - t, \quad y(0) = 0.5.$$

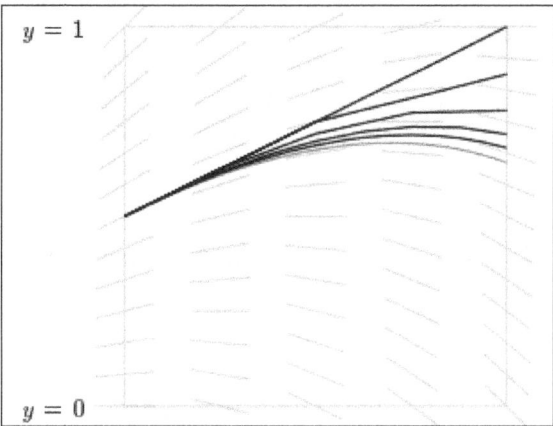

It looks very much as though you halve the error as we halve the step size. You can verify this in more detail by comparing estimates for y(1) with the true value which you know to be (t + 1)-0.5e^t at t = 1 which is equal to 2-0.5e = 0.640859.

N	estimated y(1)	true y(1)	error
2	0.875000	0.640859	0.234141
4	0.779297	...	0.138438
8	0.717108	...	0.076249
16	0.681036	...	0.040177
32	0.661505	...	0.020646
64	0.651328	...	0.010468
128	0.646130	...	0.005271
256	0.643504	...	0.002645
512	0.642184	...	0.001325
10 24	0.641522	0.640859	0.000663

This example leads to a guess which turns out to be true:

- The error in Euler's method for estimating y(t) with a given differential equation and initial condition of y at a fixed value of t is proportional to the step size chosen.

The effect of this is to make Euler's method extremely inefficient for serious calculation. It requires an enormous amount of calculation to achieve reasonable accuracy. Very roughly speaking, the amount of work it takes to achieve a certain level of accuracy is proportional to the accuracy you want. Thus if it takes N steps to get 1 digit of accuracy (answer within 1/10), it will take 10 N to get an extra valid digit (within 1/100), 100N to get 3 valid digits and 100000 N to get 6.

The reason for the way error and step size interact for Euler's method is not hard to understand at least informally. Euler's method uses the estimate

$$y(t + \Delta t) = y(t) + \Delta t \cdot y'(t) + \text{terms of order} \, (\Delta t)^2.$$

That means that the error in making each step is essentially of order (Dt)². But for a fixed interval across which you want to calculate, the number of steps necessary is proportional to the inverse of Dt, which makes it plausible that the overall error is of order (1/Dt)(Dt)²=Dt.

Illustrating Errors Graphically

You can use this to estimate the error in Euler's method even when we don't know the exact answer. For example, suppose you look at the equation,

$$y' = xy + 1, \quad y(0) = 1.$$

This is a linear equation, but if you were to use a standard formula for the solution you would see it produces an integral you cannot find explicitly. Here is the estimate we get for y(1) with various values of N:

N	y(1)
4	2.65515137
8	2.83657060
16	2.94189329
32	2.99898700
64	3.02876204
128	3.04397338
256	3.05166223
512	3.05552774
1024	3.05746579

And here is a graph of the pairs (Δ, y(1)).

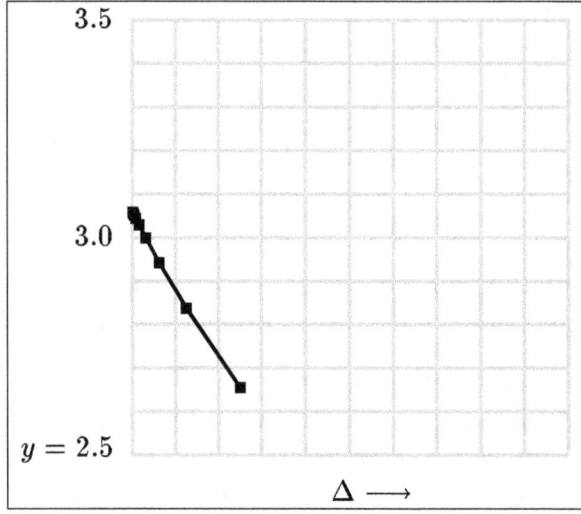

A plot of y(1) versus step size.

What you can see from this plot is that the points $(\Delta, y(1))$ are to a good approximation on a straight line. This is in complete agreement with the principle I stated above, that the error is proportional to Δ. But you can also see that to some extent you can predict what the end of the line is. This is because we can estimate from each successive pair what the slope of the purported straight line is ought to be about.

$$
\begin{array}{c|ccccc}
0 & 0 & 0 & 0 & 0 \\
2\tilde{a} & \gamma & \gamma & 0 & 0 \\
1-2\tilde{a} & \dfrac{-20\tilde{a}^2+10\tilde{a}-1}{4\gamma} & \dfrac{(2\tilde{a}-1)(4\tilde{a}-1)}{4\gamma} & \gamma & 0 \\
1 & \dfrac{24\tilde{a}^3-36\tilde{a}^2+12\tilde{a}-1}{12\tilde{a}(1-2\tilde{a})} & \dfrac{12\tilde{a}^2-6\tilde{a}+1}{12\tilde{a}(1-4\tilde{a})} & \dfrac{6\tilde{a}^2-6\tilde{a}+1}{3(4\tilde{a}-1)(2\tilde{a}-1)} & \gamma \\
\hline
 & \dfrac{24\tilde{a}^3-36\tilde{a}^2+12\tilde{a}-1}{12\tilde{a}(1-2\tilde{a})} & \dfrac{12\tilde{a}^2-6\tilde{a}+1}{12\tilde{a}(1-4\tilde{a})} & \dfrac{6\tilde{a}^2-6\tilde{a}+1}{3(4\tilde{a}-1)(2\tilde{a}-1)} & \gamma
\end{array}
$$

for each N. Of course this is only an approximation, so you don't expect any of these estimates to be exact. Here is what you get in the runs above:

N	$\Delta_N=1/N$	$y_N(1)$	$y_N(1)-y_{N-1}(1)$	estimated slope
4	0.25000000	2.65515137	*	*
8	0.12500000	2.83657060	0.18141923	-1.45135383
16	0.06250000	2.94189329	0.10532269	-1.82699868
32	0.03125000	2.99898700	0.05709371	-1.82699868
64	0.01562500	3.02876204	0.02977504	-1.90560247
128	0.00781250	3.04397338	0.01521134	-1.94705210
256	0.00390625	3.05166223	0.00768885	-1.96834562
512	0.00195312	3.05552774	0.00386551	-1.97913877
1024	0.00097656	3.05746579	0.00193806	-1.98457249

The rough proportionality certainly shows up here, and the last column tells us roughly what the constant of proportion is:

- One run of Euler's method gives you no idea of the error involved.

- Two runs with values y_1 and y_2 for step sizes D_1 and D_2 give you the estimate,

$$
\left[\frac{y_1-y_2}{\Delta_1-\Delta_2}\right]\Delta
$$

for the error with step size Δ.

Runge-Kutta Methods

To obtain numerical approximations to the solution of the initial value problem,

$$y'(x) = f(x, y(x)), \ y(x_0) = y_0,$$

at a sequence of points $x_1, x_2, \ldots$, the Euler method can be used. This method computes in turn $y_1 \approx y(x_1), y_2 \approx y(x_2), \ldots$, using the formula,

$$y(x_0 + h/2) \text{ or } y(x_0 + h).$$

The Euler method is often inefficient because of the large number of steps required to achieve a specified accuracy. Furthermore, round-off error accumulation, when many steps are used, can make the numerical results unusable. The reason for poor accuracy is that the Euler method is based on the underlying quadrature approximation

$$\int_{x_0}^{x_1} y'(x) dx \approx (x_1 - x_0) y'(x_0),$$

which is accurate only when $y(x)$ is a polynomial of degree 1. Replacing this by one of the formula,

$$\int_{x_0}^{x_1} y'(x) dx \approx (x_1 - x_0) y'((x_0 + x_1)/2,$$

or $\int_{x_0}^{x_1} y'(x) dx \approx (x_1 - x_0)(y'(x_0) + y'(x_1))/2,$ or other formulae which are exact for second degree polynomials, is the essence of the methods proposed by Runge. To use these approximations, a second "stage" has to be evaluated. That is $y'(x_0)$ is first evaluated so that $y(x_0 + h/2)$ or $y(x_0 + h)$ can then be approximated in preparation for the approximate evaluation of $y'(x_0 + h/2)$ or $y'(x_0 + h)$ respectively. Taking these ideas further, the immediate aim was to obtain methods which approximate $y_1 \approx y(x_0 + h)$ so that the error in this first and typical step can be bounded by a multiple of h^{p+1}, where the integer p, known as the order, takes on increasingly high values.

A method with s stages can be represented by a tableau, often called the "Butcher tableau",

0	0	0	0	$\cdots$	0
c_2	a_{21}	0	0	$\cdots$	0
c_3	a_{31}	a_{32}	0	$\cdots$	0
$\vdots$	$\vdots$	$\vdots$	$\vdots$		$\vdots$
c_s	a_{s1}	a_{s2}	a_{s3}	$\cdots$	0
	b_1	b_2	b_3	$\cdots$	b_s

indicating that the output approximation is equal to $y_1 = y_0 + h\sum_{i=1}^{s} b_i F_i$, where F_i is equal to $f(x_0 + hc_i, y_0 + h\sum_{j=1}^{s} a_{ij} F_j)$. For example, the two methods of order 2 already introduced above from the work of Runge, have tableaux given respectively by,

$$
\begin{array}{c|cc}
0 & & \\
1/2 & 1/2 & \\
\hline
& 0 & 1
\end{array}
\qquad ,\qquad
\begin{array}{c|cc}
0 & & \\
1 & 1 & \\
\hline
& 1/2 & 1/2
\end{array}
$$

where zero values on and above the diagonal have been omitted.

The work of Runge was extended by, who completed a discussion of order 3 methods and pointed the way to order 4, and by who gave a complete classification of order 4 methods. The most well-known method, due to Runge, has order 4 and is defined by the tableau.

$$
\begin{array}{c|cccc}
0 & 1/2 & & & \\
1/2 & 0 & 1/2 & & \\
1/2 & 0 & 0 & 1 & \\
\hline
1 & 1/6 & 1/3 & 1/3 & 1/6
\end{array}
$$

Order Conditions

The standard differential equation $y'(x) = f(x, y(x))$, $y(x_0) = y_0$, can be replaced by the autonomous system $y'(x) = f(y(x))$, as long as it is understood that the problem is multi-dimensional. This greatly simplifies the analysis and also draws attention to a short-coming of analyses based on $y'(x) = f(x, y(x))$, $y(x_0) = y_0$, if is assumed that y is a scalar-valued variable. This short-coming is that for $p > 4$, it is possible to achieve order p for a single non-autonomous problem but some lower order for a general autonomous problem. A partial explanation of this can be seen from Table.

Table: Number of conditions to achieve a specified order.

order p	1	2	3	4	5	6
Conditions for high-dimensional problem	1	2	4	8	17	37
Conditions for one-dimensional problem	1	2	4	8	16	31

which compares the number of order conditions for each problem. The number of conditions for the high-dimensional general problem to have order p is equal to the number of rooted-trees with less than or equal to p vertices.

Explicit Runge-kutta Methods

Although it is not known, for arbitrary orders, how many stages are required to achieve this order, the result is known up to order 8 and is given in Table. Also shown for comparison is the number of free parameters in an s stage method.

Table: Number of stages to achieve a specified order.

order p	1	2	3	4		5	6		7		8
Conditions	1	2	4	8		17	37		85		200
Stages	1	2	3	4	5	6	7	8	9	10	11
Parameters	1	3	6	10	15	21	28	36	45	55	66

The 8 conditions for order 4 in the case of a 4 stage method are given in equations below, for lower orders, where fewer stages are required, the conditions are identical in form but some terms are omitted from the left-hand sides. For example for a 3 stage method, terms with a factor b_4 are omitted. We see immediately that equation $b_4 a_{43} a_{32} c_2 = 1/24$ then becomes contradictory, thus confirming that 4 stages are required for order 4.

$$b_1 + b_2 + b_3 + b_4 = 1,$$

$$b_2 c_2 + b_3 c_3 + b_4 c_4 = 1/2,$$

$$b_2 c_2^2 + b_3 c_3^2 + b_4 c_4^2 = 1/3,$$

$$b_3 a_{32} c_2 + b_4 a_{42} c_2 + b_4 a_{43} c_3 = 1/6,$$

$$b_2 c_2^3 + b_3 c_3^3 + b_4 c_4^3 = 1/4,$$

$$b_3 c_3 a_{32} c_2 + b_4 c_4 a_{42} c_2 + b_4 c_4 a_{43} c_3 = 1/8,$$

$$b_3 a_{32} c_2^2 + b_4 a_{42} c_2^2 + b_4 a_{43} c_3^2 = 1/12,$$

$$b_4 a_{43} a_{32} c_2 = 1/24.$$

The actual task of solving the order conditions to derive specific methods becomes increasingly more complicated as the order increases. However, for low orders, a simple procedure exists and consists of three steps (i) choose appropriate values of $c_2, \dots, c_s$, (ii) choose $b_1, \dots, b_s$ as solutions to the equations, where applicable, amongst equations $b_1 + b_2 + b_3 + b_4 = 1$, $b_2 c_2 + b_3 c_3 + b_4 c_4 = 1/2$, $b_2 c_2^2 + b_3 c_3^2 + b_4 c_4^2 = 1/3$, $b_2 c_2^3 + b_3 c_3^3 + b_4 c_4^3 = 1/4$, (iii) solve for the a_{ij} from the remaining equations which involve these quantities linearly. For orders less than 4 these steps are all that is required. However, for order 4, the remaining equation $b_4 a_{43} a_{32} c_2 = 1/24$. still needs to be satisfied and this imposes a condition on the c values selected in step (i). This condition is that $c_4 = 1$.

Runge-kutta Pairs

In modern software for the solution of initial value problems, the stepsize is allowed to vary, taking account of estimates of the error produced in each step. To achieve this, it is standard practice to build method pairs, based on the same stages which produce an output answer of order p and a second approximation of order q. Assuming that $q > p$, the difference of these two approximations will give an asymptotically correct estimate of the error in the output value. Because, for small h, the actual local error is approximately proportional to hp^{+1}, the stepsize in the following step can be chosen to give a value close to that specified as a user tolerance. As order increases, it becomes increasingly costly, in terms of required stages, to build pairs with $q > p$ and the value $q = p - 1$ is often chosen. Even though the actual error is not estimated, the difference in the two approximations, which now behaves asymptotically like hq^{+1} can still be used to give satisfactory control.

Implicit Runge-kutta Methods

In principle, it is possible to include non-zero coefficients on and above the diagonal in a tableau. For example, the following method has order 4:

$$
\begin{array}{c|cc}
1/2-\sqrt{3}/6 & 1/4 & 1/4-\sqrt{3}/6 \\
1/2+\sqrt{3}/6 & 1/4+\sqrt{3}/6 & 1/4 \\
\hline
 & 1/2 & 1/2
\end{array}
$$

You need to ask what it means to carry out a computational step using this type of "implicit" method. Extending the formulae from explicit methods, it seems that we need to satisfy a sequence of (non-linear) equations for Y_i and F_i, $i=1,2,\ldots,s$. Once these have been solved, the values of the Fi are substituted into the equation $y_n = y_{n-1} + h\sum_{i=1}^{s} b_i F_i$ to evaluate y_n from an input approximation y_{n-1}. You note that $c_i = 1/2 \mp \sqrt{3}/6$ are the zeros of the second degree Legendre polynomial on the interval [0,1]. This method is actually a generalization, to differential equations, of the 2-point Gauss-Legendre quadrature formula. It is remarkable that a similar implicit Runge-kutta formula exists with order $2s$ for every positive integer s.

Other families of implicit Runge-kutta methods exist and many of these have been the subject of systematic study. In particular, the Radau and Lobatto families of methods sacrifice one or two, in terms of order, but have other advantages. For the so-called DIRK methods, also known as SDIRK or semi-explicit or semi-implicit methods, A has a lower triangular structure with $a_{ii} = \gamma$, $i=1,2,\ldots,s$ where the constant γ is chosen for stability reasons. In cases in which the output value $y_{n-1} + h\sum_{j=1}^{s} b_j F_j$ is identical with the final stage $y_{n-1} + h\sum_{j=1}^{s} a_{ij} F_j$, it is possible that a_{11} is equal to 0 rather than to γ, without taking away from the essential nature of a DIRK method. An example of such a method is given by the tableau

Tableau:

$$
\begin{array}{c|cccc}
0 & 0 & 0 & 0 & 0 \\
2\tilde{a} & \gamma & \gamma & 0 & 0 \\
1-2\tilde{a} & \dfrac{-20\tilde{a}^2+10\tilde{a}-1}{4\gamma} & \dfrac{(2\tilde{a}-1)(4\tilde{a}-1)}{4\gamma} & \gamma & 0 \\
1 & \dfrac{24\tilde{a}^3-36\tilde{a}^2+12\tilde{a}-1}{12\tilde{a}(1-2\tilde{a})} & \dfrac{12\tilde{a}^2-6\tilde{a}+1}{12\tilde{a}(1-4\tilde{a})} & \dfrac{6\tilde{a}^2-6\tilde{a}+1}{3(4\tilde{a}-1)(2\tilde{a}-1)} & \gamma \\
\hline
 & \dfrac{24\tilde{a}^3-36\tilde{a}^2+12\tilde{a}-1}{12\tilde{a}(1-2\tilde{a})} & \dfrac{12\tilde{a}^2-6\tilde{a}+1}{12\tilde{a}(1-4\tilde{a})} & \dfrac{6\tilde{a}^2-6\tilde{a}+1}{3(4\tilde{a}-1)(2\tilde{a}-1)} & \gamma
\end{array}
$$

where $\lambda \approx 0.4358665215$ is a zero of the polynomial $6\tilde{a}^3 - 18\tilde{a}^2 + 9\tilde{a} - 1$.

A final remark about this DIRK method. Even though formally there are 4 stages, the first of these is identical to the last stage in the *previous* step. Hence, in a large number of steps there are effectively only 3 stages per step.

Permissions

Index

A

Arbitrary Constant, 16, 33, 35-36, 38, 59, 74, 95-96

B

Bernoulli Differential Equation, 83
Boundary Value Problem, 177-178

C

Calculus, 34-35, 98, 179, 182
Cauchy Problem, 162, 164, 166
Chain Rule, 1, 3-4, 8-9, 20, 35, 37-38, 58-59, 167-171, 188, 191-192
Characteristic Curves, 159, 163-164, 172-175
Characteristic Equation, 99-103, 106-108, 112, 132-135, 143-147, 149, 164, 172, 190, 192
Complex Numbers, 27, 41, 44, 107, 133, 142
Complex Roots, 108, 133-134
Conjugate Roots, 133
Constant Coefficients, 97-98, 111, 132-133, 141, 151-153, 155
Constant Function, 34-35, 97
Cosines, 77, 111, 114, 119, 125
Cubic Polynomial, 115, 135

D

Dependent Variable, 8, 51-52, 152, 185
Derivative Function, 2
Differential Equation, 1, 16, 24-33, 38-44, 47-48, 51-81, 83-84, 86-88, 90-95, 101, 106, 116, 119, 122, 126, 129-140, 146, 152, 159, 164, 172, 176, 187, 193, 201, 208, 212, 219
Direction Field, 39
Directional Derivatives, 10, 12

E

Eigenfunction Expansion, 187
Eikonal Equation, 177
Euler-lagrange Equations, 182-183
Exponential Functions, 1, 106, 111, 121, 142

F

First-order Linear Pde, 170
Fourier Transform, 193

G

General Solution, 16, 28-29, 35, 38-39, 41, 51, 53, 55, 74-76, 83, 91, 93-105, 108-111, 113, 125, 137, 139, 142, 149, 152, 156, 162, 165, 169, 172, 176, 187, 193, 198, 200, 203

H

Hamilton-jacobi Equation, 177, 182, 185
Harmonic Waves, 196
Homogeneous Equations, 50, 92, 111

I

Imaginary Unit, 106
Indefinite Integral, 15-16, 21
Independent Variable, 27, 30, 41-42, 48, 56, 68, 151-152
Initial Value Problem, 28, 103-106, 109, 161, 165, 172, 185, 218
Initial-boundary Conditions, 157-158
Integral Curves, 39, 58, 160
Integrating Factor, 33, 55, 62-63, 65, 69-73, 75, 77-80, 83-84, 86-88, 171
Integration, 15-16, 18-20, 23, 33, 41, 50, 59-61, 65, 75-76, 81, 84, 96, 141-142, 150, 169, 185
Interval Of Validity, 28, 41-49, 84-89

L

Laplaces Equation, 190
Lax Pair, 157
Linear Function, 8, 188
Liouville Formula, 92, 95-96

N

Non-linear Differential Equation, 26, 83
Non-resonance Case, 143
Nonhomogeneous Equation, 109-111, 127, 142, 144-149
Nonlinear Equations, 55, 179
Nonzero Function, 64, 70

O

Ordinary Differentiation, 4

P

Parametric Form, 174-176, 178
Partial Differentiation, 4, 157
Particular Solution, 39, 92-96, 109-118, 121-127, 130-131, 138, 140, 142-148, 185-186
Poisson's Equation, 157
Polynomial Equation, 31
Potential Function, 55, 57-62, 65-67, 72
Product Rule, 3, 73, 77, 81

Q

Quadratic Formula, 43

Quadratic Polynomial, 120

Quasi-linear Equations, 178

Quasilinear Problem, 164

R

Real Numbers, 18, 27, 34, 87-88, 106-107, 133, 202

Real Variable, 192-193

Resonance Case, 143-144, 146

S

Scaling Behavior, 50

Sines, 77, 111, 114, 119, 125

Single Constant, 41, 118

Space Curves, 164

Square Root, 44

Steady State, 185

Superposition Principle, 143, 147

T

Tangent Line, 4-6, 166-167, 214

Tangent Plane, 8, 14, 159, 179

Tangent Vector, 13, 171-172

Taylor Expansion, 107

Trigonometric Functions, 1, 106

U

Undetermined Coefficients, 111-112, 125, 142-143

V

Variation Of Parameters, 111, 125, 127, 130, 142

Vector Fields, 160

Vector Tangent, 159

W

Wave Equation, 157, 167-168, 185, 193-199

Z

Zero Order Derivative, 99